AF360736

Enzyme Catalysis: Advances, Techniques, and Outlooks

Enzyme Catalysis: Advances, Techniques, and Outlooks

Editor

In Jung Kim

MDPI • Basel • Beijing • Wuhan • Barcelona • Belgrade • Manchester • Tokyo • Cluj • Tianjin

Editor
In Jung Kim
Kyungpook National
University
Korea

Editorial Office
MDPI
St. Alban-Anlage 66
4052 Basel, Switzerland

This is a reprint of articles from the Special Issue published online in the open access journal *Applied Sciences* (ISSN 2076-3417) (available at: https://www.mdpi.com/journal/applsci/special_issues/enzyme_catalysis_advances).

For citation purposes, cite each article independently as indicated on the article page online and as indicated below:

LastName, A.A.; LastName, B.B.; LastName, C.C. Article Title. *Journal Name* **Year**, *Volume Number*, Page Range.

ISBN 978-3-0365-5087-9 (Hbk)
ISBN 978-3-0365-5088-6 (PDF)

Contents

About the Editor . . . . **vii**

In Jung Kim
Enzyme Catalysis: Advances, Techniques, and Outlooks
Reprinted from: *Appl. Sci.* **2022**, *12*, 8036, doi:10.3390/app12168036 . . . . **1**

Thi Huong Ha Nguyen, Soo-Jin Yeom and Chul-Ho Yun
Production of a Human Metabolite of Atorvastatin by Bacterial CYP102A1 Peroxygenase
Reprinted from: *Appl. Sci.* **2021**, *11*, 603, doi:10.3390/app11020603 . . . . **5**

Joana M. C. Fernandes, Albino A. Dias and Rui M. F. Bezerra
Kinetic Analysis Misinterpretations Due to the Occurrence of Enzyme Inhibition by Reaction
Product: Comparison between Initial Velocities and Reaction Time Course Methodologies
Reprinted from: *Appl. Sci.* **2022**, *12*, 102, doi:10.3390/app12010102 . . . . **19**

Wei Zhang, Jintao Wu, Jing Xiao, Mingyao Zhu and Haichuan Yang
Compatibility and Washing Performance of Compound Protease Detergent
Reprinted from: *Appl. Sci.* **2022**, *12*, 150, doi:10.3390/app12010150 . . . . **27**

Ki Hyun Nam
Glucose Isomerase: Functions, Structures, and Applications
Reprinted from: *Appl. Sci.* **2022**, *12*, 428, doi:10.3390/app12010428 . . . . **43**

In Jung Kim and Kyoung Heon Kim
L-Fucose Synthesis Using a Halo- and Thermophilic L-Fucose Isomerase from Polyextremophilic
Halothermothrix orenii
Reprinted from: *Appl. Sci.* **2022**, *12*, 4029, doi:10.3390/app12084029 . . . . **55**

Ki Hyun Nam
Crystal Structure of Human Lysozyme Complexed with *N*-Acetyl-α-D-glucosamine
Reprinted from: *Appl. Sci.* **2022**, *12*, 4363, doi:10.3390/app12094363 . . . . **67**

About the Editor

In Jung Kim

Dr. In Jung Kim performed her Ph.D. research in Food Bioscience and Technology at the University of Korea in South Korea (2008–2014) on a project about enzymatic saccharification of lignocellulosic biomass. After a 5-year postdoctoral research at the University of Korea (2015–2019) on a project about biosynthesis of rare sugars and their source, she did her overseas postdoctoral research in at the University of Greifswald in Germany (2020–2021) to perform the project on biotechnological production of aroma compounds. Currently, she is a research professor at Kyungpook National University in Korea.

applied sciences

Editorial

Enzyme Catalysis: Advances, Techniques, and Outlooks

In Jung Kim

Division of Plant Biosciences, School of Applied Biosciences, College of Agriculture and Life Science, Kyungpook National University, Daegu 41566, Korea; ij0308@knu.ac.kr; Tel.: +82-53-950-6654

Citation: Kim, I.J. Enzyme Catalysis: Advances, Techniques, and Outlooks. *Appl. Sci.* **2022**, *12*, 8036. https://doi.org/10.3390/app12168036

Received: 4 August 2022
Accepted: 9 August 2022
Published: 11 August 2022

Publisher's Note: MDPI stays neutral with regard to jurisdictional claims in published maps and institutional affiliations.

Biocatalysis using enzymes is a powerful strategy that can be employed in a variety of industries for the production of biofuel, biochemicals, pharmaceuticals, and foods, etc. [1]. Compared with the chemical method, enzyme catalyst is considered as an eco-friendly and more selective tool, facilitating a sustainable and circular economy. Enzymes possessing high catalytic efficiency, stability to harsh conditions (i.e., heat, extreme pH, and high salinity), and desired substrate specificity are considered to be process-compatible. To discover such robust enzymes, various technologies of bioinformatics, machine learning, protein engineering, and high-throughput screening (HTS) have been developed [2]. With the aid of those cutting edge technologies, it is more readily achievable to discover and develop novel and superior enzymes. Noteworthy, Frances Arnold's award of the Nobel Prize in 2018 for protein engineering facilitates the scaled-up applications of enzyme in industries.

This Special Issue "Enzyme Catalysis: Advances, Techniques, and Outlooks" aimed to contribute to the advancement of related fields by covering broad subjects and technological developments on industrially relevant enzymes, such as sugar isomerase (i.e., gluose/fucose isomerase, lysozyme, protease, and cytochrome P450. Under this Special Issue, a total of six papers were published, comprising five original research papers and one review paper.

Cytochrome P450 is considered as industrially useful biocatalyst due to its versatility [3]. Nguyen et al. reported an efficient enzymatic strategy using bacterial CYP102A1 peroxidase for the synthesis of 4-OH atorvastatin from atorvastatin, which is the key human metabolite having a beneficial effect on cardiovascular disease and hyperlipidemia [4]. Specifically, a superior mutant of CYP102A1 was obtained with a catalytic efficiency of 1.8 min^{-1} with an aid of HTS and enzyme characterization. This approach was conducted in the presence of hydrogen peroxide but without NADPH, which was further discussed on its advantage over NADPH-requiring monooxygenase.

The Michaelis–Menten equation is routinely used for the investigation of enzyme kinetics. However, when the reaction product acts as the inhibitor, this conventional model does not necessarily reflect the enzyme's actual behaviors, causing a misinterpretation. For reliability, Fernandes et al. developed a robust methodology based on an integrated Michaelis–Menten equation (IMME) [5], which could be helpful for the investigation of enzymes such as the well-known β-glucosidase, often encountering end-product inhibition (i.e., glucose) [6].

Zhang et al. developed a more efficient detergent than commercial ones by preparing for in-house blends composed of different proteases (i.e., alkaline protease, keratinase, and trypsin) combined with detergent additives including surfactants, anti-redeposition agents, and water-softening agents [7]. During the process, various approaches, such as whiteness determination, microscope scanning, Fourier transform infrared spectroscopy, and X-ray photoelectron spectroscopy were used to assess the cleaning efficiency.

Sugar isomerase is a class of enzymes that mediate the interconversion of sugar isomers in a reversible manner. With its broad substrate scope, this enzyme has wide applications in food/beverage, pharmaceutical/medicinal, and beauty industries [8].

The well-recognized glucose isomerase (GI), which is involved in catalyzing the reaction between D-glucose and D-fructose, and D-xylose and D-xylulose has already been commercially available for the production of high-fructose corn syrup. Recognizing the significant role of the enzyme both in academic and industrial sectors, Nam presented an overview of the functions, structure, and applications of GI with a focus on structural insight into the metal-binding site and interaction with the inhibitor [9].

Extremozymes are capable of maintaining their functions even in harsh conditions, such as high temperature/salinity and extreme pH levels, and are thus considered desirable for industrial application. In this regard, Kim et al. reported a unique L-fucose isomerase from the polyextremophile *Halothermothrix orenii* (HoFucI) that exhibits both halo- and thermophilicities when synthesizing L-fucose from L-fuculose. Such a robustness offers a great advantage for the L-fucose isomerase-mediated production of rare sugars [10].

Lysozyme, a glycoside hydrolase that breaks down the peptidoglycan, the major cell wall component of microorganisms, is a natural antimicrobial agent. Nam reported three forms of crystal structure of human lysozyme: one native and two *N*-acetyl-α-D-glucosamine-complexed configurations, providing an insight into the mechanism by which lysozyme recognizes sugar molecules when involved in the immune responses to microbial infections [11].

Overall, this Special Issue detailed wide applications of industrially useful enzymes with excellent experimental data and improved their methodology in addition to providing state-of-the art reviews on their technological advances. I, as the guest editor, am sincerely grateful for all the efforts from the authors, reviewers, editors, and staff of the editorial office of Applied Sciences for their work in successfully completing this Special Issue.

Funding: This work was funded by the National Research Foundation of Korea (NRF) (NRF-2020R1A6A3A03039153 and NRF-2022R1I1A1A01072158).

Conflicts of Interest: The authors declare no conflict of interest.

References

1. Wu, S.; Snajdrova, R.; Moore, J.C.; Baldenius, K.; Bornscheuer, U.T. Biocatalysis: Enzymatic synthesis for industrial applications. *Angew. Chem. Int. Ed.* **2021**, *60*, 88–119. [CrossRef] [PubMed]
2. Yi, D.; Bayer, T.; Badenhorst, C.P.S.; Wu, S.; Doerr, M.; Höhne, M.; Bornscheuer, U.T. Recent trends in biocatalysis. *Chem. Soc. Rev.* **2021**, *50*, 8003–8049. [CrossRef] [PubMed]
3. Grobe, S.; Badenhorst, C.P.S.; Bayer, T.; Hamnevik, E.; Wu, S.; Grathwol, C.W.; Link, A.; Koban, S.; Brundiek, H.; Großjohann, B.; et al. Engineering regioselectivity of a P450 monooxygenase enables the synthesis of ursodeoxycholic acid via 7β-hydroxylation of lithocholic acid. *Angew. Chem. Int. Ed.* **2021**, *60*, 753–757. [CrossRef] [PubMed]
4. Nguyen, T.H.; Yeom, S.-J.; Yun, C.-H. Production of a human metabolite of atorvastatin by bacterial CYP102A1 peroxygenase. *Appl. Sci.* **2021**, *11*, 603. [CrossRef]
5. Fernandes, J.M.C.; Dias, A.A.; Bezerra, R.M.F. Kinetic analysis misinterpretations due to the occurrence of enzyme inhibition by reaction product: Comparison between initial velocities and reaction time course methodologies. *Appl. Sci.* **2022**, *12*, 102. [CrossRef]
6. Kim, I.J.; Bornscheuer, U.T.; Nam, K.H. Biochemical and structural analysis of a glucose-tolerant β-glucosidase from the hemicellulose-degrading *Thermoanaerobacterium saccharolyticum*. *Molecules* **2022**, *27*, 290. [CrossRef] [PubMed]
7. Zhang, W.; Wu, J.; Xiao, J.; Zhu, M.; Yang, H. Compatibility and washing performance of compound protease detergent. *Appl. Sci.* **2022**, *12*, 150. [CrossRef]
8. Kim, I.J.; Kim, D.H.; Nam, K.H.; Kim, K.H. Enzymatic synthesis of L-fucose from L-fuculose using a fucose isomerase from *Raoultella* sp. and the biochemical and structural analyses of the enzyme. *Biotechnol. Biofuels* **2019**, *12*, 282. [CrossRef] [PubMed]
9. Nam, K.H. Glucose isomerase: Functions, structures, and applications. *Appl. Sci.* **2022**, *12*, 428. [CrossRef]
10. Kim, I.J.; Kim, K.H. L-Fucose synthesis using a halo- and thermophilic L-fucose isomerase from polyextremophilic *Halothermothrix orenii*. *Appl. Sci.* **2022**, *12*, 4029. [CrossRef]
11. Nam, K.H. Crystal structure of human lysozyme complexed with *N*-acetyl-α-D-glucosamine. *Appl. Sci.* **2022**, *12*, 4363. [CrossRef]

Short Biography of Author

Dr. In Jung Kim performed her Ph.D. research in Food Bioscience and Technology at the University of Korea in South Korea (2008–2014) on a project about enzymatic saccharification of lignocellulosic biomass. After a 5-year postdoctoral research at the University of Korea (2015–2019) on a project about biosynthesis of rare sugars and their source, she did her overseas postdoctoral research in at the University of Greifswald in Germany (2020–2021) to perform the project on biotechnological production of aroma compounds. Currently, she is a research professor at Kyungpook National University in Korea.

Article

Production of a Human Metabolite of Atorvastatin by Bacterial CYP102A1 Peroxygenase

Thi Huong Ha Nguyen [1,2], Soo-Jin Yeom [1,2] and Chul-Ho Yun [1,2,*]

1   School of Biological Sciences and Biotechnology, Graduate School, Chonnam National University, Yongbong-ro 77, Gwangju 61186, Korea; 177597@jnu.ac.kr (T.H.H.N.); soojin258@chonnam.ac.kr (S.-J.Y.)
2   School of Biological Sciences and Technology, Chonnam National University, Yongbong-ro 77, Gwangju 61186, Korea
*   Correspondence: chyun@jnu.ac.kr; Tel.: +82-62-530-2194

Featured Application: The peroxygenase activity of CYP102A1 can be used to prepare atorvastatin drug metabolites and possibly other drugs for biotechnological applications.

Abstract: Atorvastatin is a widely used statin drug that prevents cardiovascular disease and treats hyperlipidemia. The major metabolites in humans are 2-OH and 4-OH atorvastatin, which are active metabolites known to show highly inhibiting effects on 3-hydroxy-3-methylglutaryl-CoA reductase activity. Producing the hydroxylated metabolites by biocatalysts using enzymes and whole-cell biotransformation is more desirable than chemical synthesis. It is more eco-friendly and can increase the yield of desired products. In this study, we have found an enzymatic strategy of P450 enzymes for highly efficient synthesis of the 4-OH atorvastatin, which is an expensive commercial product, by using bacterial CYP102A1 peroxygenase activity with hydrogen peroxide without NADPH. We obtained a set of CYP102A1 mutants with high catalytic activity toward atorvastatin using enzyme library generation, high-throughput screening of highly active mutants, and enzymatic characterization of the mutants. In the hydrogen peroxide supported reactions, a mutant, with nine changed amino acid residues compared to a wild-type among tested mutants, showed the highest catalytic activity of atorvastatin 4-hydroxylation ($1.8\ \mathrm{min}^{-1}$). This result shows that CYP102A1 can catalyze atorvastatin 4-hydroxylation by peroxide-dependent oxidation with high catalytic activity. The advantages of CYP102A1 peroxygenase activity over NADPH-supported monooxygenase activity are discussed. Taken together, we suggest that the P450 peroxygenase activity can be used to produce drugs' metabolites for further studies of their efficacy and safety.

Keywords: CYP102A1; atorvastatin; 4-hydroxy atorvastatin; hydrogen peroxide; P450 peroxygenase; NADPH

Citation: Nguyen, T.H.H.; Yeom, S.-J.; Yun, C.-H. Production of a Human Metabolite of Atorvastatin by Bacterial CYP102A1 Peroxygenase. *Appl. Sci.* **2021**, *11*, 603. https://doi.org/10.3390/app11020603

Received: 30 November 2020
Accepted: 7 January 2021
Published: 10 January 2021

1. Introduction

Atorvastatin is a cholesterol-lowering drug widely used in treating hypercholesterolemia and preventing cardiovascular disease [1,2]. This statin drug inhibits 3-hydroxy-3-methyl-glutaryl-coenzyme A (HMG-CoA) reductase activity, which catalyzes the rate-limiting step in cholesterol biosynthesis [3]. The atorvastatin can be hydroxylated to 2-OH and 4-OH atorvastatin by cytochrome P450 (CYP or P450) enzyme, CYP3A4, in the human liver [4] (Figure 1, Figure S1). Along with this, atorvastatin's hydroxylated metabolites have been reported as active metabolites. Among statin drugs, only the atorvastatin has metabolites that can inhibit HMG-CoA reductase equipotent to that of the parent drug [5]. About 70% of the HMG-CoA reductase inhibition achieved with atorvastatin has been observed with its 2-OH and 4-OH atorvastatin [6]. In addition, these two active metabolites have higher efficacy in lowering LDL (low-density lipoprotein) cholesterol when compared with other statins [7]. However, systematic studies have not been performed on the safety, efficacy, and toxicity of these metabolites, although the metabolites can contribute to atorvastatin's cholesterol-lowering effect.

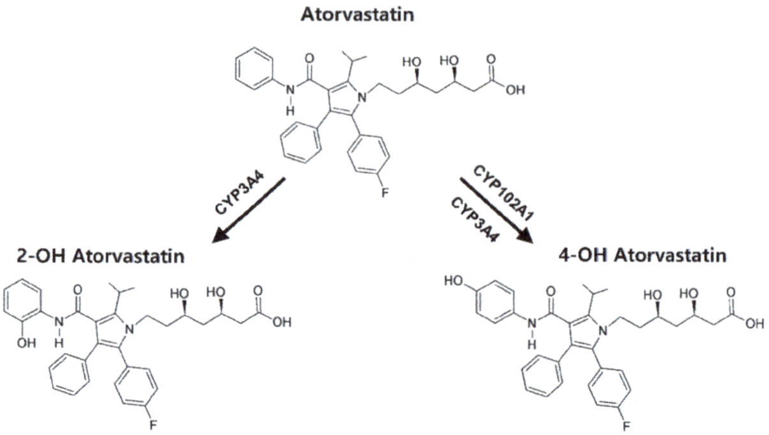

Figure 1. Chemical structures of atorvastatin and its metabolites, 2-OH and 4-OH atorvastatin. The metabolites are made in the human liver by CYP3A4. Bacterial CYP102A1 mutants can make only one metabolite, 4-OH atorvastatin.

During the drug development process, which is guided by the U.S. Food and Drug Administration (FDA), the drug metabolites in safety testing (MIST) should be separated to evaluate the safety testing and drug toxicity [8,9]. Although chemical synthesis methods can prepare some drug metabolites of concern, other metabolites such as chiral compounds may not be easily prepared by these methods. Those metabolites can be synthesized by P450 enzymes, including human P450s [10]. Particularly, the CYP102A1 (P450 BM3) from *Bacillus megaterium* demonstrated as a possible biocatalyst for producing human metabolites, which could be useful in fine-chemical biosynthesis and the pharmaceutical fields [11,12]. The CYP102A1 mutants, with new or enhanced catalytic activities toward drugs, could be created by site-directed mutagenesis, random mutagenesis, and rational design [13]. The CYP102A1 was also proposed for industrial application as a potential monooxygenase with high catalytic activities and stabilities [11,14].

The P450s' typical function is to catalyze the oxidation of a wide range of physiological substrates and foreign chemicals. P450s reductively activate molecular oxygen (O_2) by using two electrons transferred to P450s by NAD(P)H via P450 reductase (CPR) [15]. Otherwise, hydrogen peroxide (H_2O_2) can be used as an oxidant for oxygen activation of P450s via the peroxide shunt pathway [16]. Using peroxides is interesting for monooxygenation reactions catalyzed by P450s as the reaction does not require an electron transfer partner and cofactor NAD(P)H. The low-cost H_2O_2 can lead to using the P450 peroxygenase in industrial-scale synthesis [17].

Reports found that some of the CYP102A1 mutants obtained via random mutagenesis could catalyze regioselective hydroxylation of atorvastatin to produce 4-OH atorvastatin, which is one of two of the major metabolites in humans, by NADPH-dependent CYP102A1 catalyzed reaction [12] (Figure 1). In this work, we examined the peroxygenase activity of CYP102A1 mutants in having high atorvastatin 4-hydroxylation activity supported by H_2O_2 and compared this to the mutants' activities supported by the NADPH regeneration system. We found that the peroxygenase activity of CYP102A1 can be used as a biocatalyst to catalyze the reaction of atorvastatin hydroxylation because CYP102A1 peroxygenase with a low cost has higher activity than that of the NADPH-supported activity.

2. Materials and Methods

2.1. Materials

Atorvastatin, hydrogen peroxide, and glucose oxidase were purchased from Sigma-Aldrich (St. Louis, MO, USA). 2-OH atorvastatin and 4-OH atorvastatin were obtained from Toronto Research Chemicals (North York, ON, Canada) Other chemicals and solvents were purchased with the highest grade from commercial suppliers.

2.2. Hydroxylation of Atorvastatin by CYP102A1

The CYP102A1 mutants' catalytic activity in the hydroxylation of atorvastatin was supported by NADPH regeneration system or hydrogen peroxide. The reaction mixtures include 200 µM of atorvastatin and 0.20 µM CYP102A1 in 100 mM of a potassium phosphate buffer (pH 7.4), and were pre-incubated for 5 min at 37 °C. Hydroxylation of atorvastatin with an NADPH regeneration system (0.5 mM $NADP^+$, 10 mM glucose-6-phosphate, and 1.0 IU yeast glucose-6-dehydrogenase/mL) was used to initiate reaction. The reaction mixtures were incubated at 37 °C for 30 min and stopped by 600 µL of ice-cold ethyl acetate.

The hydroxylation of atorvastatin was supported by hydrogen peroxide; an aliquot of 10 mM H_2O_2 initiated the reactions. The reaction mixtures were incubated at 37 °C for 10 min and stopped by 600 µL of ice-cold ethyl acetate. Quercetin (50 µM) was added as an internal standard to this solution. Then, the mixtures were vortexed for 3 min. After centrifugation (3000× g, 10 min), aliquots (350 µL) of the ethyl acetate layer were transferred to a clean glass tube and the ethyl acetate was removed under a gentle stream of N_2 gas. Each sample was dissolved in 170 µL of 30% of HPLC mobile phase (see below) and 30 µL was used for the quantitation of the samples.

The atorvastatin's metabolites were analyzed by HPLC, using a Gemini C18 column (4.6 × 150 mm, 5 µm; Phenomenex, Torrance, CA, USA) with the mobile phase A (79.9% water, 20% acetonitrile, and 0.1% formic acid) and the mobile phase B (9.9% water, 90% acetonitrile and 0.1% formic acid). The elution rate was 1.0 mL/min by a gradient pump (LC-20AD; Shimadzu, Kyoto, Japan) with the following gradient: 0–8 min, 30% mobile phase B; 8–13 min, gradually increased to 60% mobile phase B; 13–20 min decreased to 30% mobile phase B; and 20–30 min fixed at 30% mobile phase B. The UV at 260 nm detected the eluent. The CYP102A1 mutants used in this study are shown in Supplementary Table S1.

Calibration standards of atorvastatin and quercetin were constructed from a blank sample (a reaction mixture without substrate and internal standard), nine samples of atorvastatin (2–500 µM), and thirteen samples of quercetin covering 0.2–500 µM (Figure S2). The peak area ratio of atorvastatin to internal standard (quercetin) was linear with respect to the analyte concentration over the range of 0.2–500 µM. When 2.4 volume of ethyl acetate was used to extract atorvastatin and internal standard in buffer solution, the extraction efficiencies of atorvastatin and internal standard were 79% and 74%, respectively, at the concentrations used in the assay. Quantitation of the metabolites was done by comparing the peak areas of each metabolite to mean peak areas of the internal standard (50 µM). The detection limit of 4-OH atorvastatin was 0.20 µM.

2.3. Construction of an Expression Vector for the Heme Domain of CYP102A1 Mutant

The polymerase chain reaction (PCR) was performed using a Thermal Cycler (BioRad, Richmond, CA, USA) and Phusion high-fidelity DNA polymerase (Thermo Fisher Scientific, Waltham, MA, USA). For the expression of the heme domain of M371, the gene-coding heme domain (1–472 residues) was amplified by the PCR with oligonucleotide primers BamHI/XbaI restriction sites: BM3-BamHI forward, 5′-AGCGGATCCATGACAATTAAAGAAATGCC-3′ and BM3-XbaI reverse, 5′-TCTAGACTATTAGCGTACTTTTTTAGCAGACTG-3′. The PCR product was resolved on a 1% (*v*/*v*) agarose gel, purified, digested with BamHI and XbaI and ligated into a pCWOri plasmid [18,19]. The recombinant pCW vector containing M371-heme was expressed in *Escherichia coli* DH5αF′IQ.

2.4. Expression of CYP102A1 Mutants

The plasmids of whole M371 [20,21] and M371-heme were transformed into *E. coli* DH5α F'-IQ cells and spread on Luria-Bertani (LB) agar plate containing ampicillin (100 μg/mL). The single colony was grown in 5 mL of LB medium containing ampicillin (100 μg/mL) at 37 °C and 170 rpm for 10 h. The cell culture's (1% *v/v*) aliquots were inoculated in 200 mL of a Terrific Broth (TB) medium with 100 μg/mL ampicillin, 1 mM thiamine, trace elements, 50 μM $FeCl_3$, 1 mM $MgCl_2$, and 2.5 mM $(NH_4)_2SO_4$. The cells were grown at 37 °C and 170 rpm to an OD_{600} of 0.6–0.8. Then, isopropyl-β-D-thiogalactopyranoside (0.50 mM) and δ-aminolevulinic acid (1.0 mM) were added for CYP protein expression. After the cells were grown at 30 °C and 170 rpm for 22 h, the cells were collected by centrifugation (15 min, 5000× *g*, 4 °C). The cell pellet was resuspended in 100 mM Tris-HCl (pH 7.6) containing 0.50 mM ethylenediaminetetraacetic acid (EDTA) and 500 mM sucrose, and lysed by sonication (Sonicator, Heat Systems—Ultrasonics, Plainview, NJ, USA). After the lysate was centrifuged at 100,000× *g* (90 min, 4 °C), P450 concentrations were examined from the CO-difference spectra using the extinction molecular coefficient: $\varepsilon = 91$ mM^{-1} · cm^{-1} [22]. The supernatant of the lysate was used for the catalytic activity assay of the CYP102A1 enzymes.

2.5. LC-MS Analysis

To characterize the major metabolite of atorvastatin produced by the CYP102A1, the reaction mixtures (1 mL) included 0.20 μM whole M371 enzyme and 200 μM atorvastatin in 100 mM potassium phosphate buffer (pH 7.4). The initial reaction was started by the NADPH-generating system at 37 °C for 30 min. The atorvastatin and metabolites were analyzed using Shimadzu LCMS-2010 EV system (Shimadzu, Kyoto, Japan). The metabolites and substrates were separated on a Shim-pack VP-ODS column (2.0 mm i.d. 250 mm; Shimadzu) using a mobile phase with acetonitrile–water–formic acid (70:30:0.1, *v/v/v*) at a flow rate of 0.16 mL/min. To identify the metabolites, electrospray ionization in positive mode recorded the mass spectra. The interface and detector voltages were 4.4 and 1.5 kV, respectively. The nebulization gas flow was set at 1.5 L/min. The interface, curve desolvation line, and heat block temperatures were 250 °C, 230 °C, and 200 °C, respectively [23].

2.6. Atorvastatin Hydroxylation's Kinetic Parameters and Total Turnover Numbers

For determining the kinetic parameters of CYP102A1 mutants, the reaction mixtures included an atorvastatin substrate at a concentration of 5–200 μM in 100 mM potassium phosphate buffer (pH 7.4) and 0.20 μM CYP102A1. The reaction mixtures were preincubated for 5 min at 37 °C. An aliquot of 10 mM H_2O_2 initiated the reactions, which were performed for 10 min at 37 °C.

To determine total turnover numbers (TTNs) of CYP102A1 mutants, the reaction mixtures included atorvastatin substrate at a concentration of 200 μM in 100 mM potassium phosphate buffer (pH 7.4) and 0.20 μM CYP102A1. The reaction mixtures were preincubated for 5 min at 37 °C. An aliquot of 10 mM H_2O_2 initiated the reactions, which were performed for 30 s, 1, 2, 3, 4, 5, 10, 20, and 30 min at 37 °C.

The kinetic parameters of CYP102A1 catalyzed reactions supported by NADPH were determined by the reaction, including 5–500 μM of atorvastatin in 100 mM potassium phosphate buffer (pH 7.4) and 0.20 μM enzymes. The NADPH-generating systems were added to the initial reaction and the reaction mixtures were incubated at 37 °C for 30 min. The products were extracted with ethyl acetate and analyzed by HPLC as described above. The kinetic parameters were analyzed by GraphPad Prism software (GraphPad Software, San Diego, CA, USA).

2.7. Comparison of Atorvastatin Hydroxylation Activity Supported by External Addition of Hydrogen Peroxide and In Situ Hydrogen Peroxide Generation

For the reactions with externally added hydrogen peroxide, the reaction mixtures included 2 mL total volume including 200 μM atorvastatin in a potassium phosphate buffer (100 mM, pH 7.4) and 0.20 μM CYP102A1. The reaction mixtures were pre-incubated for 5 min at 37 °C. An aliquot of 10 mM H_2O_2 initiated the reactions.

For the reactions with in situ hydrogen peroxide production, the reaction mixtures included 2 mL total volume including 200 μM atorvastatin, 0.20 μM CYP102A1,4 g/L of glucose, and 10 U/mL of glucose oxidase in a potassium phosphate buffer (100 mM, pH 7.4).

The reactions were performed at 37 °C at 2, 5, 10, 20, 30, 60, and 120 min point time, aliquots of 250 μL reaction mixture were taken and the reactions were stopped by adding 600 μL of ice-cold ethyl acetate for 4-OH atorvastatin measurement.

2.8. Measuring Hydrogen Peroxide Concentration

The hydrogen peroxide concentration was measured by a spectrophotometric assay [24]. The reactions included in situ hydrogen peroxide production, and the reaction mixtures with 2 mL included 200 μM atorvastatin, 0.20 μM CYP102A1, 4 g/L of glucose, and 10 U/mL of glucose oxidase in a potassium phosphate buffer (100 mM, pH 7.4). After the reactions were performed at 37 °C during 2–60 min, aliquots of 100 μL reaction mixture were taken for the hydrogen peroxide measurement. The assay of 1 mL total volume including 100 μL of H_2O_2 production reaction was mixed with 0.28 mM 2,2′-azinobis(3-ethylbenzthiazoline-6-sulfonate) (ABTS) and 0.5 U/mL of the horseradish peroxidase (HPR) in 100 mM potassium phosphate buffer (pH 7.4). The formation of the green ABTS cation radical was measured spectrophotometrically for 1 min at 420 nm ($\varepsilon = 36\ \mathrm{mM}^{-1} \cdot \mathrm{cm}^{-1}$) [25].

2.9. The Effect of Cosolvent on Hydroxylation of Atorvastatin Supported by Hydrogen Peroxide

The reaction mixtures included the atorvastatin substrate at a concentration of 200 μM, 0.20 μM CYP102A1, and cosolvent (methanol, ethanol, isopropanol, or acetonitrile glycerol) of 0.5 to 10% (*v/v*) in a potassium phosphate buffer (100 mM, pH 7.4). The reaction mixtures were pre-incubated for 5 min at 37 °C. An aliquot of 10 mM H_2O_2 initiated the reactions, which were performed for 10 min at 37 °C.

2.10. Spectral Binding Titration

The spectral binding titrations of substrates to the CYP102A1 were determined with a Shimadzu 1601PC Spectrometer at 23 °C, as described previously [26]. The atorvastatin's binding affinity to four CYP102A1 mutants was determined by titrating 1.0 μM enzyme with the substrate in a total of 1 mL volume in a potassium phosphate buffer (100 mM, pH 7.4). The absorption of the UV–vis spectral difference from 350 nm to 500 nm was recorded after each substrate addition and plotted against the added substrate concentration (0–100 μM). The spectrally determined dissociation constants (K_d values) were determined using GraphPad Prism software (GraphPad Software, San Diego, CA, USA).

2.11. Statistical Analysis

All experiments were performed three times. The values are presented as means with a standard error of mean (SEM).

3. Results and Discussion

3.1. Hydroxylation of Atorvastatin

At first, we screened the atorvastatin hydroxylation activities of a set of CYP102A1 mutants, which showed high activities toward several drugs supported by NADPH [20,21]. We found some mutants showed higher catalytic activity of atorvastatin 4-hydroxylation when compared to those of our previous work [12]. The mutants M179 (0.89 min^{-1}),

M221 (1.25 min^{-1}), M371 (2.0 min^{-1}), and M387 (0.83 min^{-1}) showed 3.2–7.8-fold higher formation rate of 4-OH atorvastatin than the template M16V2 to make an enzyme library by error-prone PCR (0.26 min^{-1}) (Figure 2). Based on these results, M16V2, M179, M221, M371, and M387 were selected and used for studying atorvastatin hydroxylation activities supported by H_2O_2. Each mutant bore amino acid changes compared to WT CYP102A1, as described in Table S1. Another major metabolite of atorvastatin, 2-OH product, in the human liver, was not found in the CYP102A1-catalyzed reaction.

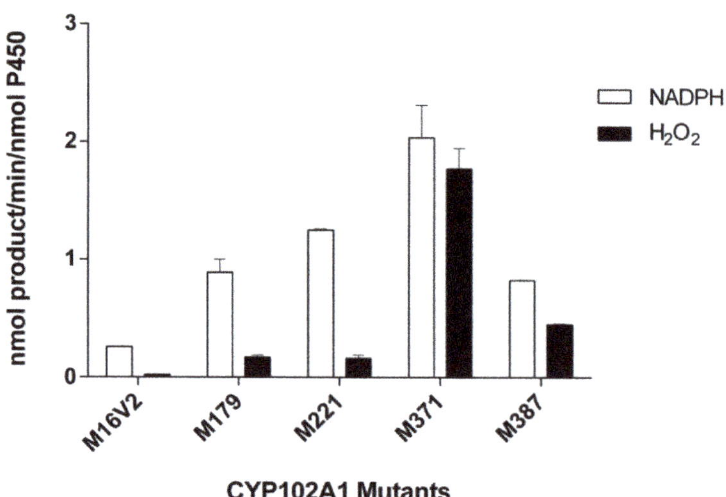

Figure 2. Activity of atorvastatin 4-hydroxylation catalyzed by CYP102A1 mutants. The reaction mixtures included atorvastatin substrate (200 µM) and 0.20 µM of CYP102A1 in 100 mM of a potassium phosphate buffer (pH 7.4). For NADPH-supported reactions, an NADPH-generating system was used to initiate the reaction, which was incubated for 30 min at 37 °C. For H_2O_2 supported reactions, 10 mM H_2O_2 was used to initiate the reaction, which was incubated for 10 min at 37 °C.

The mutants M179, M221, M371, and M387 showed 7.5-fold, 7.1-fold, 76.6-fold, and 19.5-fold higher rates of 4-OH atorvastatin formation than M16V2, respectively, when the reaction was performed with the externally added H_2O_2 (10 mM). We found that 10 mM was the most optimal H_2O_2 concentration to support the peroxygenase activity of CYP102A1 (Figure S3). The mutant M371 (1.8 min^{-1}) showed the highest catalytic activity toward atorvastatin among the tested mutants (Figure 2).

M371's catalytic activities toward atorvastatin did not show large differences between NADPH- and H_2O_2 supported reactions (Figure S4). Based on these results, M371 showed the highest product formation rates in both the NADPH and H_2O_2 dependent reactions. Thus, we selected M371 for constructing the heme domain in an expression vector. The major metabolites and substrates were characterized by the HPLC (Figure 3) and LC-MS (Figure S5) results.

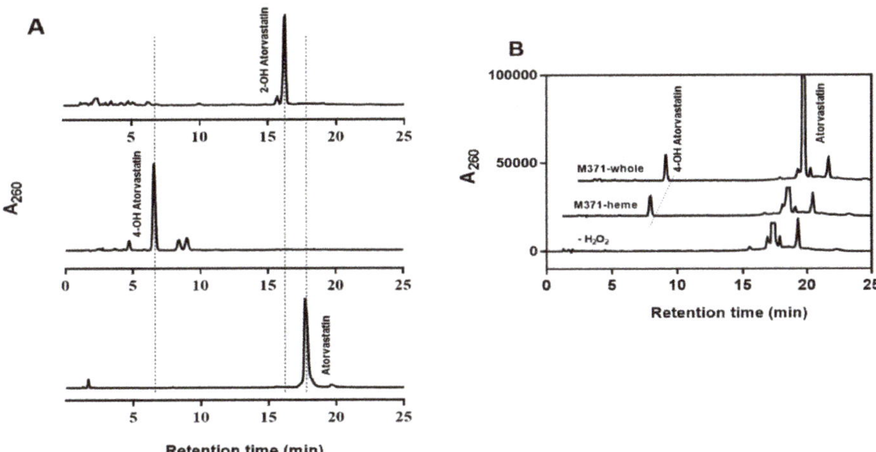

Figure 3. (**A**) HPLC chromatograms of substrate atorvastatin and its metabolites, 2-OH and 4-OH products. The concentration of each compound was 200 µM. (**B**) HPLC chromatograms of atorvastatin metabolites produced by whole M371 and M371-heme domain in the presence of H_2O_2. The hydroxylation of 200 µM atorvastatin by 0.20 µM CYP102A1 (M371 whole enzyme or heme domain) in 100 mM of a potassium phosphate buffer (pH 7.4) was supported by externally added hydrogen peroxide (10 mM) at 37 °C for 10 min. The reaction mixtures' peaks of HPLC chromatograms were identified by comparing their retention times to those of the following standards: authentic 4-OH atorvastatin (t_R = 6.8 min), 2-OH atorvastatin (t_R = 16.2 min), and atorvastatin (t_R = 17.3 min).

3.2. The Kinetic Parameters and TTNs of Atorvastatin Hydroxylation Reactions Supported by Hydrogen Peroxide

The mutant M371, which has both domains of heme and reductase (so-called whole M371), and M371-heme domain were used to determine the kinetic parameters of atorvastatin 4-hydroxylation (Figure 4, Table 1). The k_{cat} value of whole M371 (4.3 min^{-1}) increased 3.3-fold when compared to that of the M371 heme domain (1.3 min^{-1}). The K_m values of M371 and M371-heme were 52 µM and 106 µM, respectively. The M371-heme domain showed a 2-fold increased K_m value when compared with M371. The catalytic efficiency (k_{cat}/K_m) of 4-OH atorvastatin formation by whole M371 and M471-heme domain were 0.083 and 0.012 ($min^{-1} \cdot \mu M^{-1}$), respectively. These results showed that whole M371 was more efficient for 4-OH atorvastatin formation than M371-heme (6.9-fold).

Table 1. Kinetic parameters of atorvastatin 4-hydroxylation by whole M371 and M371-heme domain.

CYP102A1	k_{cat} (min^{-1})	K_m (µM)	k_{cat}/K_m ($min^{-1} \cdot \mu M^{-1}$)
whole M371	4.3 ± 0.3	52 ± 11	0.083 ± 0.018
M371-heme domain	1.3 ± 0.2	106 ± 35	0.012 ± 0.004

Whole M371 and M371-heme were used to determine the TTNs of atorvastatin 4-hydroxylation supported by H_2O_2 at the reaction times of 30 s, 1, 2, 3, 4, 5, 10, 20, and 30 min (Figure 5). The overall product formation was in the range of 3.2–17.3 TTNs (Figure 5). The whole M371 showed a higher 4-OH atorvastatin formation rate than that of M371-heme during all of the indicated reaction times. In addition, the results showed that 4-OH atorvastatin formation reached a plateau after 10 min of a reaction.

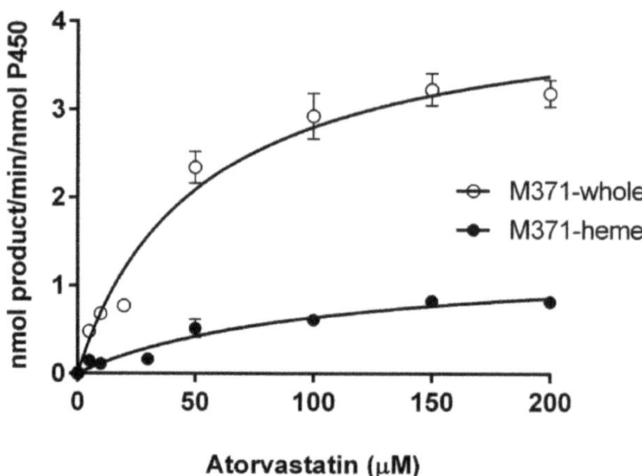

Figure 4. Kinetic parameters of atorvastatin 4-hydroxylation by whole M371 and M371-heme in the presence of H_2O_2. The reactions include 5–200 μM atorvastatin substrate and 0.20 μM of CYP102A1 M371 whole enzyme (○) or heme domain in 100 mM of a potassium phosphate buffer (pH 7.4) (●). We added 10 mM H_2O_2 to initiate the reaction, which was incubated for 5 min at 37 °C.

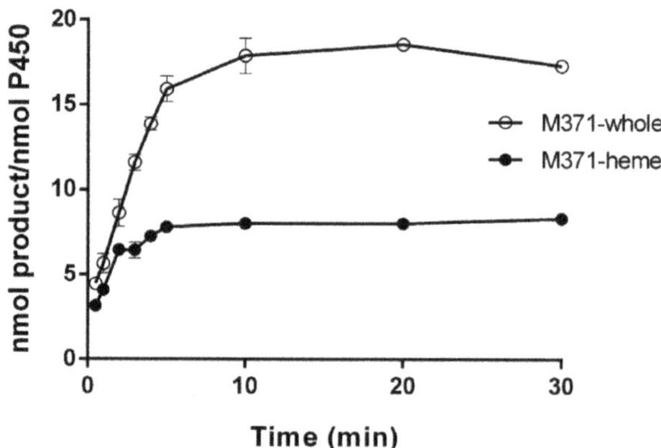

Figure 5. Total turnover numbers of atorvastatin 4-hydroxylation by CYP102A1 supported by H_2O_2. The reaction mixtures included atorvastatin substrate at concentration of 200 μM and 0.20 μM of whole M371 (○) or heme-domain in 100 mM of a potassium phosphate buffer (pH 7.4) (●). We added 10 mM H_2O_2 to initiate the reaction and the reaction mixtures were incubated for 30 s, 1, 2, 3, 4, 5, 10, 20, and 30 min at 37 °C.

Taken together, the results of kinetic parameters and TTNs with whole enzyme and heme-domain of M371 indicate that the whole protein shows higher catalytic activity than the heme-domain, even if the reductase domain of the whole enzyme does not involve in the peroxygenase activity of the heme-domain. The catalytic activity of M371-heme toward atorvastatin is much lower (~45%) than the whole M371 (Figure 5). This result might be due to a more stable conformation of whole M371 enzyme than that of M371-heme enzyme. This result suggests the reductase domain makes the heme-domain a more functional conformation for its peroxygenase activity.

When we compared the kinetic parameters of whole M371 catalyzed 4-hydroxylation of atrovastatin supported by NADPH and H_2O_2, the k_{cat} value of H_2O_2 supported reaction was 1.4-fold higher than that of the NADPH-supported reaction (Table 1; Table S2). The

K_m value of the H_2O_2-supported reaction was 2.8-fold lower than that of the NADPH-supported reaction. This result means that the H_2O_2-supported reaction showed a higher catalytic efficiency (3.3-fold) than that of the NADPH-supported reaction.

3.3. Comparison of Atorvastatin 4-Hydroxylation Activity of CYP102A1 Supported by External Addition and In Situ Generation of Hydrogen Peroxide

A continuous low-level supply or in situ generation of hydrogen peroxide is important for peroxygenase's stability [27,28]. The overall productivity can be increased due to enhanced enzyme stability at tailored H_2O_2 generation rates. Here, we compared the peroxygenase activity of CYP102A1 supported by externally added H_2O_2 (10 mM) or by in situ generation of H_2O_2 via glucose and glucose oxidase.

The catalytic activity of atorvastatin 4-hydroxylation supported by externally adding 10 mM H_2O_2 increased rapidly from a 2–10 min reaction time, after that, 4-OH formation rates reached a plateau (Figure 6). After 30 min, the reaction rate decreased slightly.

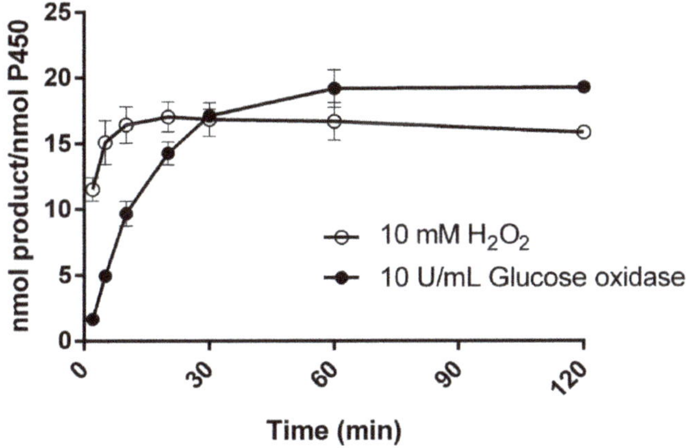

Figure 6. Comparison of external addition of H_2O_2 generation and in situ H_2O_2 generation to support the atorvastatin 4-hydroxylation activity of CYP102A1. The reaction mixtures included atorvastatin substrate at a concentration of 200 μM and 0.20 μM CYP102A1 in 100 mM of a potassium phosphate buffer (pH 7.4). To start the reaction, 10 mM H_2O_2 was externally added to the reaction mixtures (○) or 4 g/L of glucose and 10 U/mL of glucose oxidase were added (●).

When glucose oxidase and glucose were used to produce H_2O_2 continuously, the atorvastatin 4-hydroxylation activity gradually increased from a 2–30 min reaction time. The product formation reached a plateau at 30 min and then increased slightly up to 60 min. The results show that in situ H_2O_2 generation is more suitable for 4-OH atorvastatin formation than that of externally adding H_2O_2 to support the peroxygenase activity of CYP102A1.

When the concentrations of H_2O_2 with glucose oxidase and glucose were determined under the same experimental conditions with M371, the H_2O_2 concentrations increased to 2.7 mM after 5 min, was constant until 10 min, and then decreased to 1.3 mM after 60 min (Figure S6). These results show the distinctive effects of the two H_2O_2 systems, the external addition and in situ generation, were observed on the atorvastatin 4-hydroxylation activity of CYP102A1.

3.4. Effect of Cosolvent on Atorvastatin Hydroxylation Activity Supported by Hydrogen Peroxide

We examined the cosolvent's effect on the peroxygenase activity of CYP102A1 toward atorvastatin to find a good solvent system for the atorvastatin hydroxylation activity. Although methanol, ethanol, isopropanol, and acetonitrile showed the 4-OH atorvastatin formation rate generally decreasing (Figure 7A), glycerol gradually increased the 4-OH atorvastatin formation rate when its concentrations increased up to 2.0% (*v/v*) (Figure 7B).

When the glycerol concentration increased by more than 3%, the activity gradually decreased. The results show the peroxygenase activity of the whole M371 toward atorvastatin reached the highest (5.1 min^{-1}) when 2% glycerol was added in the reaction mixtures.

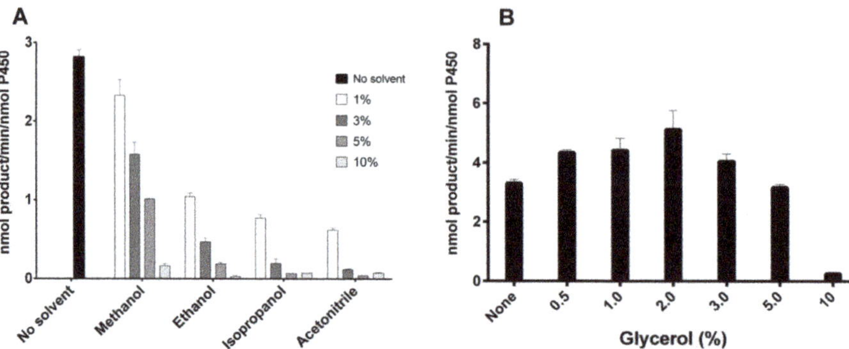

Figure 7. The cosolvent's effect on atorvastatin hydroxylation supported by hydrogen peroxide. The reaction mixtures included atorvastatin substrate at 200 µM, 0.20 µM whole M371, and indicated cosolvent in 100 mM of a potassium phosphate buffer (pH 7.4). (**A**) Cosolvent (methanol, ethanol, isopropanol, and acetonitrile) with 1%, 3%, 5%, and 10% (*v/v*). (**B**) Cosolvent: glycerol with 0.5%, 1%, 2%, 3%, 5%, and 10% (*v/v*).

3.5. Spectral Binding Titration

To examine the differences in the substrate atorvastatin's binding affinity to mutants M179, M221, M371, and M387, spectral binding titration was performed (Figure 8). Binding of atorvastatin to all of the mutants tested here produced a pronounced Type II spectral shift, with a decrease at 390 nm and an increase at 420 nm, showing an increase in the heme-domain's low-spin fraction. The mutant's K_d values ranged between 2.5 and 3.7 µM, although M221 showed a higher K_d value (17 µM) than others. When 4-OH atorvastatin, the product, was added to M371, a modified Type I spectral shift was produced with a decrease at 420 nm and an increase at 390 nm (Figure S7). The K_d value of 4-OH atorvastatin to whole M371 was 6.2 µM.

Although deuterated analogs were not used as internal standards to the analysis of atorvastatin and its metabolite by LC-MS, the kinetic findings in this study should be reliable because commercially obtained 4-OH atorvastatin was also used to confirm the results of subsequent kinetic studies. We found the calibration methods used to quantify the metabolite using quercetin as an internal standard were similar to the direct comparison of external 4-OH atorvastatin standard. The extraction efficiency of 4-OH atorvastatin was 75% under the experimental conditions used in this study.

Drug metabolites in humans produced by P450s are critical for evaluating drug efficacy and safety in the drug discovery and development process. Lower activities of P450 enzymes are suggested for the H_2O_2-supported reaction rather than the NADPH-supported reaction, which can be explained by the lack of general acid–base residues in the P450 active sites [17]. This study showed that several CYP102A1 mutants can catalyze human drug metabolites with high peroxygenase activity to produce the human metabolite 4-OH atorvastatin without requiring NADPH, an expensive cofactor [29]. Our results suggest that the catalytic activities of CYP102A1 peroxygenase toward atorvastatin can be improved by enzyme engineering via random mutagenesis and site-directed mutation. The CYP102A1 mutants selected by high-throughput screening [12,20,21] showed high catalytic activity toward atorvastatin. Atorvastatin inhibits HMG-CoA reductase and is a very popular clinical drug used for hypertension and hyperlipidemia. The 4-OH atorvastatin is known as an active metabolite, which is an expensive commercial product. By further improving CYP102A1 peroxygenase activity and stability, the peroxygenase activity can be used to prepare drug metabolites of atorvastatin and possibly other drugs for industrial purposes.

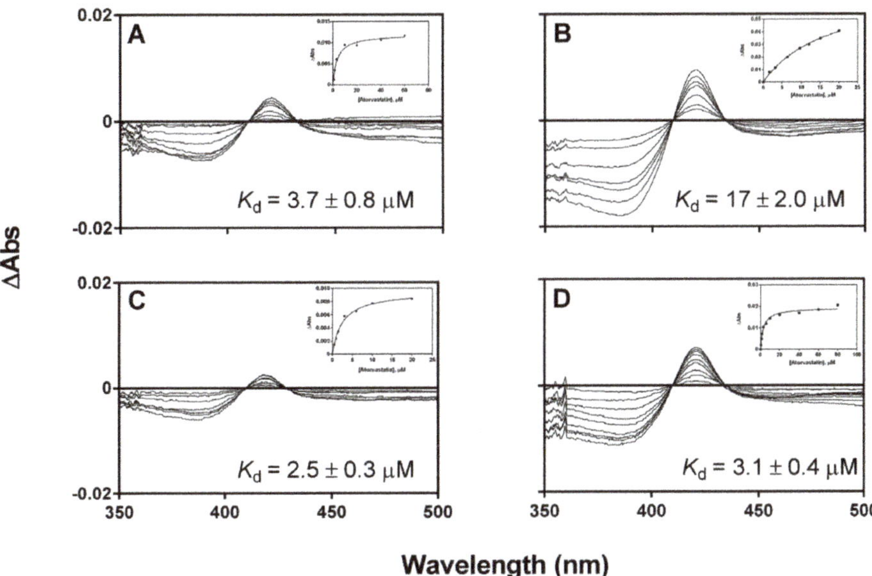

Figure 8. Binding titration of atorvastatin to the CYP102A1. The samples for binding titration contained 0–100 μM atorvastatin in 100 mM of a potassium phosphate buffer (pH 7.4) and 1 μM of M179 (**A**), M221 (**B**), M371 (**C**), and M387 (**D**). Spectrally determined dissociation constants (K_d) were also shown.

4. Conclusions

In this study, we found a simple enzymatic strategy of CYP102A1 enzymes for a high synthesis efficiency for the 4-OH atorvastatin by using bacterial CYP102A1 peroxygenase activity with hydrogen peroxide. A set of CYP102A1 mutants with high catalytic activity toward atorvastatin were obtained using enzyme library generation, high-throughput screening of highly active mutants, and enzymatic characterization of the mutants. In H_2O_2-supported reactions, the mutant M371 among the tested mutants showed the highest catalytic activity of atorvastatin 4-hydroxylation. Improved enzymatic properties of CYP102A1 peroxygenase activity over the NADPH-supported activity were found. This result shows that CYP102A1 peroxygenase activity can catalyze atorvastatin 4-hydroxylation by peroxide-dependent oxidation with high catalytic activity. These results suggest that the peroxygenase activity of CYP102A1 mutants can be developed to produce drugs' metabolites to further study their efficacy and safety.

Supplementary Materials: The following are available online at https://www.mdpi.com/2076-3417/11/2/603/s1, Table S1: The amino acid sequences of M16V2 and CYP102A1 mutants, Table S2: Kinetic parameters of atorvastatin 4-hydroxylation by CYP102A1 supported by NADPH and dissociation constants for atorvastatin, Figure S1: Proposed pathway for the atorvastatin metabolism in human, Figure S2: Standard curves of internal standard, quercetin, and atorvastatin, Figure S3: Dependence of atorvastatin 4-hydroxylation by CYP102A1 mutant on the H_2O_2 concentration, Figure S4. HPLC chromatograms of atorvastatin and its metabolite produced by CYP102A1 mutant with and without NADPH, Figure S5: LC/MS spectra of atorvastatin and its major metabolite produced by CYP102A1 mutant (M371), Figure S6: Measurement of H_2O_2 concentration during atorvastatin 4-hydroxylation by whole M371 via in situ H_2O_2 generation, Figure S7: Binding titration of 4-OH atorvastatin to the CYP102A1.

Author Contributions: Conceptualization, S.-J.Y. and C.-H.Y.; writing—original draft preparation, T.H.H.N. and C.-H.Y.; supervision, S.-J.Y. and C.-H.Y.; funding acquisition, C.-H.Y. All authors have read and agreed to the published version of the manuscript.

Funding: This research was funded by the National Research Foundation of Korea (NRF-2018R1A4A1023882), Republic of Korea.

Institutional Review Board Statement: Not applicable.

Informed Consent Statement: Not applicable.

Data Availability Statement: The data presented in this study are available on request from the corresponding author.

Conflicts of Interest: The authors declare that they have no conflict of interest.

References

1. Pahan, K. Lipid-lowering drugs. *Cell Mol. Life Sci.* **2006**, *63*, 1165–1178. [CrossRef] [PubMed]
2. Arca, M.; Gaspardone, A. Atorvastatin Efficacy in the Primary and Secondary Prevention of Cardiovascular Events. *Drugs* **2007**, *67*, 29–42. [CrossRef] [PubMed]
3. Taniguti, E.H.; Ferreira, Y.S.; Stupp, I.J.V.; Fraga-Junior, E.B.; Doneda, D.L.; Lopes, L.; Rios-Santos, F.; Lima, E.; Buss, Z.S.; Viola, G.G.; et al. Atorvastatin prevents lipopolysaccharide-induced depressive-like behaviour in mice. *Brain Res. Bull.* **2019**, *146*, 279–286. [CrossRef] [PubMed]
4. Park, J.-E.; Kim, K.-B.; Bae, S.K.; Moon, B.-S.; Liu, K.-H.; Shin, J.-G. Contribution of cytochrome P450 3A4 and 3A5 to the metabolism of atorvastatin. *Xenobiotica* **2008**, *38*, 1240–1251. [CrossRef] [PubMed]
5. Poli, A. Atorvastatin: Pharmacological characteristics and lipid-lowering effects. *Drugs* **2007**, *67* (Suppl. 1), 3–15. [CrossRef] [PubMed]
6. Partani, P.; Verma, S.M.; Gurule, S.; Khuroo, A.; Monif, T. Simultaneous quantitation of atorvastatin and its two active metabolites in human plasma by liquid chromatography electrospray tandem mass spectrometry. *J. Pharm. Anal.* **2014**, *4*, 26–36. [CrossRef]
7. Hoffart, E.; Ghebreghiorghis, L.; Nussler, A.; Thasler, W.; Weiss, T.; Schwab, M.; Burk, O. Effects of atorvastatin metabolites on induction of drug-metabolizing enzymes and membrane transporters through human pregnane X receptor. *Br. J. Pharmacol.* **2012**, *165*, 1595–1608. [CrossRef]
8. Guengerich, F.P. Introduction: Human Metabolites in Safety Testing (MIST) Issue. *Chem. Res. Toxicol.* **2009**, *22*, 237–238. [CrossRef]
9. Luffer-Atlas, D.; Atrakchi, A. A decade of drug metabolite safety testing: Industry and regulatory shared learning. *Expert Opin. Drug Metab. Toxicol.* **2017**, *13*, 897–900. [CrossRef]
10. Chun, Y.-J.; Shimada, T.; Waterman, M.R.; Guengerich, F.P. Understanding electron transport systems of *Streptomyces* cytochrome P450. *Biochem. Soc. Trans.* **2006**, *34*, 1183–1185. [CrossRef]
11. Yun, C.-H.; Kim, K.-H.; Kim, D.-H.; Jung, H.-C.; Pan, J.-G. The bacterial P450 BM3: A prototype for a biocatalyst with human P450 activities. *Trends Biotechnol.* **2007**, *25*, 289–298. [CrossRef] [PubMed]
12. Kang, J.-Y.; Ryu, S.H.; Park, S.-H.; Cha, G.S.; Kim, D.-H.; Kim, K.-H.; Hong, A.W.; Ahn, T.; Pan, J.-G.; Joung, Y.H.; et al. Chimeric cytochromes P450 engineered by domain swapping and random mutagenesis for producing human metabolites of drugs. *Biotechnol. Bioeng.* **2014**, *111*, 1313–1322. [CrossRef] [PubMed]
13. Whitehouse, C.J.C.; Bell, S.G.; Wong, L.-L. P450(BM3) (CYP102A1): Connecting the dots. *Chem. Soc. Rev.* **2012**, *41*, 1218–1260. [CrossRef] [PubMed]
14. Bernhardt, R. Cytochromes P450 as versatile biocatalysts. *J. Biotechnol.* **2006**, *124*, 128–145. [CrossRef] [PubMed]
15. Anari, M.R.; Josephy, P.D.; Henry, T.; O'Brien, P.J. Hydrogen peroxide supports human and rat cytochrome P450 1A2-catalyzed 2-amino-3-methylimidazo [4,5-f] quinoline bioactivation to mutagenic metabolites: Significance of cytochrome P450 peroxygenase. *Chem. Res. Toxicol.* **1997**, *10*, 582–588. [CrossRef]
16. Munro, A.W.; McLean, K.J.; Grant, J.L.; Makris, T.M. Structure and function of the cytochrome P450 peroxygenase enzymes. *Biochem. Soc. Trans.* **2018**, *46*, 183–196. [CrossRef]
17. Shoji, O.; Watanabe, Y. Peroxygenase reactions catalyzed by cytochromes P450. *J. Biol. Inorg. Chem.* **2014**, *19*, 529–539. [CrossRef]
18. Farinas, E.T.; Schwaneberg, U.; Glieder, A.; Arnold, F.H. Directed Evolution of a Cytochrome P450 Monooxygenase for Alkane Oxidation. *Adv. Synth. Catal.* **2001**, *343*, 601–606. [CrossRef]
19. Kim, D.-H.; Kim, K.-H.; Kim, D.-H.; Liu, K.-H.; Jung, H.-C.; Pan, J.-G.; Yun, C.-H. Generation of human metabolites of 7-ethoxycoumarin by bacterial cytochrome P450 BM3. *Drug Metab. Dispos.* **2008**, *36*, 2166–2170. [CrossRef]
20. Le, T.-K.; Jang, H.-H.; Nguyen, H.; Doan, T.; Lee, G.-Y.; Park, K.; Ahn, T.; Joung, Y.; Kang, H.-S.; Yun, C.-H. Highly regioselective hydroxylation of polydatin, a resveratrol glucoside, for one-step synthesis of astringin, a piceatannol glucoside, by P450 BM3. *Enzym. Microb. Technol.* **2016**, *97*, 34–42. [CrossRef]
21. Jang, H.-H.; Ryu, S.-H.; Le, T.-K.; Doan, T.T.M.; Nguyen, T.H.H.; Park, K.D.; Yim, D.-E.; Kim, D.-H.; Kang, C.-K.; Ahn, T.; et al. Regioselective C-H hydroxylation of omeprazole sulfide by *Bacillus megaterium* CYP102A1 to produce a human metabolite. *Biotechnol. Lett.* **2017**, *39*, 105–112. [CrossRef] [PubMed]

22. Omura, T.; Sato, R. The carbon monoxide-binding pigment of liver microsomes. I. Evidence for its hemoprotein nature. *J. Biol. Chem.* **1964**, *239*, 2370–2378. [CrossRef]
23. Kim, K.-H.; Kang, J.-Y.; Kim, D.; Park, S.-H.; Park, S.; Kim, D.; Park, K.; Lee, Y.J.; Jung, H.; Pan, J.-G.; et al. Generation of Human Chiral Metabolites of Simvastatin and Lovastatin by Bacterial CYP102A1 Mutants. *Drug Metab. Dispos. Biol. Fate Chem.* **2010**, *39*, 140–150. [CrossRef] [PubMed]
24. López, C.; Cavaco-Paulo, A. In-situ Enzymatic Generation of Hydrogen Peroxide for Bleaching Purposes. *Eng. Life Sci.* **2008**, *8*, 315–323. [CrossRef]
25. Sygmund, C.; Santner, P.; Krondorfer, I.; Peterbauer, C.K.; Alcalde, M.; Nyanhongo, G.S.; Guebitz, G.M.; Ludwig, R. Semi-rational engineering of cellobiose dehydrogenase for improved hydrogen peroxide production. *Microb. Cell Fact.* **2013**, *12*, 38. [CrossRef]
26. Hosea, N.A.; Miller, G.P.; Guengerich, F.P. Elucidation of Distinct Ligand Binding Sites for Cytochrome P450 3A4. *Biochemistry* **2000**, *39*, 5929–5939. [CrossRef]
27. Paul, C.E.; Churakova, E.; Maurits, E.; Girhard, M.; Urlacher, V.B.; Hollmann, F. In situ formation of H_2O_2 for P450 peroxygenases. *Bioorg. Med. Chem.* **2014**, *22*, 5692–5696. [CrossRef]
28. Freakley, S.J.; Kochius, S.; van Marwijk, J.; Fenner, C.; Lewis, R.J.; Baldenius, K.; Marais, S.S.; Opperman, D.J.; Harrison, S.T.L.; Alcalde, M.; et al. A chemo-enzymatic oxidation cascade to activate C–H bonds with in situ generated H_2O_2. *Nat. Commu.* **2019**, *10*, 4178. [CrossRef]
29. Behrendorff, J.B.Y.H.; Huang, W.; Gillam, E.M.J. Directed evolution of cytochrome P450 enzymes for biocatalysis: Exploiting the catalytic versatility of enzymes with relaxed substrate specificity. *Biochem. J.* **2015**, *467*, 1–15. [CrossRef]

applied sciences

MDPI

Article

Kinetic Analysis Misinterpretations Due to the Occurrence of Enzyme Inhibition by Reaction Product: Comparison between Initial Velocities and Reaction Time Course Methodologies

Joana M. C. Fernandes [1], Albino A. Dias [1,2] and Rui M. F. Bezerra [1,2,*]

[1] CITAB—Centre for the Research and Technology of Agro-Environmental and Biological Sciences, UTAD—Universidade de Trás-os-Montes e Alto Douro, 5000-801 Vila Real, Portugal; joanaf@utad.pt (J.M.C.F.); jdias@utad.pt (A.A.D.)

[2] Department of Biology and Environment, UTAD—Universidade de Trás-os-Montes e Alto Douro, 5000-801 Vila Real, Portugal

* Correspondence: bezerra@utad.pt; Tel.: +351-259-350-344

Featured Application: The integrated Michaelis–Menten equation (IMME) provides a viable and accurate methodology for enzyme kinetic studies when the product is also an inhibitor. This methodology is important in a wide range of applications, including in the preclinical kinetic assays of potential drug candidates. This methodology is a powerful tool for overcoming the inaccuracies associated with archaic curve linearization methods for initial velocities determination mainly when the reaction product is an enzyme inhibitor.

Abstract: The Michaelis–Menten equation (MME) has been extensively used in biochemical reactions, but it is not appropriate when the reaction product inhibits the enzyme. Under these circumstances, each determined initial velocity, v_0, is one experimental point that actually belongs to a different MME because enzymatic product inhibition occurs as the reaction starts. Furthermore, the inhibition effect is not constant, since the concentration of the product inhibitor rises as time increases. To unveil the hidden enzyme inhibition and to simultaneously demonstrate the superiority of an integrated Michaelis–Menten equation (IMME), the same range of data points, assuming product inhibition and the presence of a second different inhibitor, was used for kinetic analysis with both methodologies. This study highlights the superiority of the IMME methodology for when the enzyme is inhibited by the reaction product, giving a more coherent inhibition model and more accurate kinetic constants than the classical MME methodology.

Keywords: enzyme inhibition; integrated Michaelis–Menten equations; reaction product inhibition; two mutually exclusive inhibitors

Citation: Fernandes, J.M.C.; Dias, A.A.; Bezerra, R.M.F. Kinetic Analysis Misinterpretations Due to the Occurrence of Enzyme Inhibition by Reaction Product: Comparison between Initial Velocities and Reaction Time Course Methodologies. *Appl. Sci.* **2022**, *12*, 102. https://doi.org/10.3390/app12010102

Academic Editor: In Jung Kim

Received: 21 October 2021
Accepted: 21 December 2021
Published: 23 December 2021

Publisher's Note: MDPI stays neutral with regard to jurisdictional claims in published maps and institutional affiliations.

1. Introduction

Enzymes are catalysts with diverse substrate specificities that have been used to modulate biochemical transformations and which play an important role in a wide range of applications, including proteomics research, medical diagnosis, food processing, biofuel production, and environmental monitoring [1]. Furthermore, studies of enzyme kinetics with more than one inhibitor have been specially focused in the pharmaceutical industry on the development of more effective drugs that act at signal transduction cascades and metabolic pathway levels [2].

At the beginning of the 20th century, Henri [3] proposed the first mathematical formulation for enzymatic kinetics, corresponding to an equation that represented product versus time. Slightly later (1913), Michaelis and Menten made a proposal based on an equation of initial velocity versus substrate concentration using the catalytic constant (k_{cat}) and a substrate constant ($K_s \approx K_m$) [4], which has been the basis of the great majority of kinetic studies carried out until today. However, from a formal point of view, the occurrence of enzyme inhibition via a reaction product was not considered in this equation. This fact deserves special attention, since it has been reported that 91.7% of human enzymes exhibit inhibition and the most inhibitory interactions result from the structural similarities between substrates and inhibitors [5]. This emphasizes the extreme importance of product inhibition in kinetic studies. Since product inhibition can significantly retard the rates of enzyme-catalyzed reactions [6], the accumulated product constitutes the main obstacle for the determination of the true initial velocities using conventional assay methods [7].

When the reaction product inhibits the enzyme, the obtained initial velocity points adjust to the Michaelis–Menten equation (MME), and are always affected by the inhibitor's action [8,9]. However, kinetic studies under these conditions are usually performed by adjusting the initial velocities to an MME (without considering the effect of the inhibitor's presence). However, if the product is an inhibitor, the initial velocity determined cannot be the "true" initial velocity without the inhibitor, because the inhibitor (a reaction product) has been there since the reaction started. Thus, the utilization of an appropriate integrated Michaelis–Menten equation (IMME) provides a viable and unique methodology that can overcome such limitations. Nevertheless, before parameter estimation, model discrimination is required as a first step in order to find an IMME model that best fits the experimental points. Actually, the use of integrated equations to study enzyme kinetics is no longer a complex task since the required nonlinear regression methods are available in software spreadsheets using traditional widespread desktop tools, such as Office Excel [8–12].

The main aim of this work is to present and validate a methodology based on an IMME (with different inhibitor models) that is capable of overcoming the misinterpretations that arise when using usual initial velocities methodology [13] when an enzyme is inhibited by the reaction product. Both methodologies (MME and IMME) were used with same simulated data points (product vs. time) and the results were compared in order to find the best one.

2. Materials and Methods

2.1. Simulated Data for the Graphical Representation of Product Versus Time

Simulated data representing product versus time curves with integrated equations considering the absence of product inhibition (Equation (1), Table 1 and Figure 1a) and also competitive inhibition by the reaction product (Equation (2), Table 1 and Figure 1b) were used to obtain four graphical representations of different K_m/K_i (Figure 2). The following arbitrary parameters were used in the simulation: (a) $V_{max} = 18$ µM min^{-1}; $K_m = 500$ µM; $S_0 = 100$ µM; $K_{ic} = 0.5$ µM; (b) $V_{max} = 18.0$ µM min^{-1}; $K_m = 100$ µM; $S_0 = 100$ µM; $K_{ic} = 20$ µM; (c) $V_{max} = 18.0$ µM min^{-1}; $K_m = 10$ µM; $S_0 = 100$ µM; $K_{ic} = 2$ µM. (d) $V_{max} = 18.0$ µM min^{-1}; $K_m = 2$ µM; $S_0 = 100$ µM; $K_{ic} = 5$ µM.

Table 1. Integrated Michaelis–Menten equations obtained from the model explained in Figure 1, assuming product inhibition [10].

Kinetic Model *	Equation	
WI	$t = -\frac{1}{V_{max}}\left\{ K_m ln\frac{[S_t]}{[S_0]} + ([S_t] - [S_0]) \right\}$	(1)
CI	$t = -\frac{1}{V_{max}}\left\{ K_m\left(\frac{[S_0]}{K_{ic}} + \frac{[I_1]}{K_{ic}} + 1\right) ln\frac{[S_t]}{[S_0]} + \left(1 - \frac{K_m}{K_{ic}}\right)([S_t] - [S_0]) \right\}$	(2)
NCI	$t = -\frac{1}{V_{max}}\left\{ K_m\left(\frac{[S_0]}{K_i} + \frac{[I_1]}{K_i} + 1\right) ln\frac{[S_t]}{[S_0]} + \left(1 - \frac{K_m}{K_i} + \frac{[I_1]}{K_i} + \frac{[S_0]}{K_i}\right)([S_t] - [S_0]) - \frac{1}{2K_i}\left([S_t]^2 - [S_0]^2\right) \right\}$	(3)
UCI	$t = -\frac{1}{V_{max}}\left\{ K_m ln\frac{[S_t]}{[S_0]} + \left(1 + \frac{[S_0]}{K_{iu}} + \frac{[I_1]}{K_{iu}}\right)([S_t] - [S_0]) - \frac{1}{2K_{iu}}\left([S_t]^2 - [S_0]^2\right) \right\}$	(4)
MI	$t = -\frac{1}{V_{max}}\left\{ K_m\left(\frac{[S_0]}{K_{ic}} + \frac{[I_1]}{K_{ic}} + 1\right) ln\frac{[S_t]}{[S_0]} + \left(1 - \frac{K_m}{K_{ic}} + \frac{[S_0]}{K_{iu}} + \frac{[I_1]}{K_{iu}}\right)([S_t] - [S_0]) - \frac{1}{2K_{iu}}\left([S_t]^2 - [S_0]^2\right) \right\}$	(5)

* WI—without inhibition; CI—competitive inhibition; NCI—noncompetitive inhibition; UCI—uncompetitive inhibition; MI—mixed inhibition.

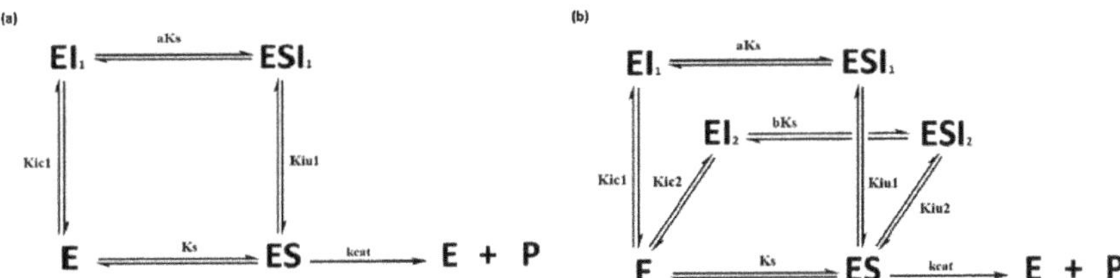

Figure 1. (**a**) Mixed linear inhibition model (MI), where: E—enzyme; ES—complex enzyme inhibitor; ESI—complex enzyme inhibitor substrate; EI—complex enzyme inhibitor; P—product; Km—Michaelis–Menten constant; K_{ic}, K_{iu}—dissociation constants; k_{cat}—catalytic constant. (**b**) Kinetic mechanism for two exclusive mixed linear inhibitors (I_1 and I_2) with K_{iu1}/K_{ic1} = a and K_{iu2}/K_{ic2} = b.

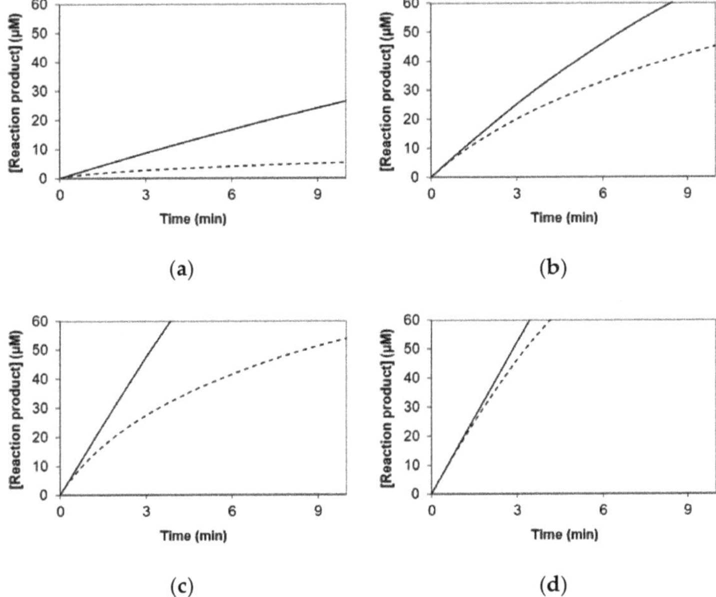

Figure 2. Theoretical simulation of product versus time curves with integrated equations assuming V_{max} = 18 μM min^{-1}; S_0 = 100 μM; and the following K_m/K_{ic} ratios: (**a**) K_m/K_{ic} = 1000; (**b**) K_m/K_{ic} = 5 (non-saturating conditions); (**c**) K_m/K_{ic} = 5 (saturating conditions); (**d**) K_m/K_{ic} = 0.4. The integrated equations utilized were without inhibition (solid line, Equation (1)) and in the presence of competitive inhibition by the reaction product (dashed line, Equation (2)).

2.2. Simulated Data for Comparison between Both Methodologies

Simulated data points [P] vs. time were obtained, taking into account the presence of mixed inhibition by the reaction product [I_1] and furthermore the presence of a second different mixed inhibitor, [I_2] = 10 μM (Table 2, Equation (9), two mixed mutually exclusive inhibitors). Data points were simulated with the following arbitrary kinetic parameters: (K_m = 500 μM; V_{max} = 0.120 μmol min^{-1} mg^{-1}; K_{ic1} = 100 μM; K_{iu1} = 50 μM; K_{ic2} = 50 μM; K_{iu2} = 5 μM) at six substrate concentrations (100; 200; 400; 800; 1000; 2000 μM).

Table 2. Integrated Michaelis–Menten equations obtained from the model explained in Figure 1, considering the simultaneous presence of two mutually exclusive inhibitors when one of which is a reaction product [8].

Kinetic Model *	Equation	
CI	$t = -\frac{1}{V_{max}}\left\{K_m\left(\frac{[S_0]}{K_{ic1}} + \frac{[I_1]}{K_{ic1}} + \frac{[I_2]}{K_{ic2}} + 1\right)ln\frac{[S_t]}{[S_0]} + \left(1 - \frac{K_m}{K_{ic1}}\right)([S_t] - [S_0])\right\}$	(6)
NCI	$t = -\frac{1}{V_{max}}\left\{K_m\left(\frac{[S_0]}{K_{i1}} + \frac{[I_1]}{K_{i1}} + \frac{[I_2]}{K_{i2}} + 1\right)ln\frac{[S_t]}{[S_0]} + \left(1 - \frac{K_m}{K_{i1}} + \frac{[I_2]}{K_{i2}} + \frac{[I_1]}{K_{i1}} + \frac{[S_0]}{K_{i1}}\right)([S_t] - [S_0]) - \frac{1}{2K_{i1}}\left([S_t]^2 - [S_0]^2\right)\right\}$	(7)
UCI	$t = -\frac{1}{V_{max}}\left\{K_m ln\frac{[S_t]}{[S_0]} + \left(1 + \frac{[S_0]}{K_{iu1}} + \frac{[I_2]}{K_{iu2}} + \frac{[I_1]}{K_{iu1}}\right)([S_t] - [S_0]) - \frac{1}{2K_{iu1}}\left([S_t]^2 - [S_0]^2\right)\right\}$	(8)
MI	$t = -\frac{1}{V_{max}}\left\{K_m\left(\frac{[S_0]}{K_{ic1}} + \frac{[I_1]}{K_{ic1}} + \frac{[I_2]}{K_{ic2}} + 1\right)ln\frac{[S_t]}{[S_0]} + \left(1 - \frac{K_m}{K_{ic1}} + \frac{[S_0]}{K_{iu1}} + \frac{[I_2]}{K_{iu2}} + \frac{[I_1]}{K_{iu1}}\right)([S_t] - [S_0]) - \frac{1}{2K_{iu1}}\left([S_t]^2 - [S_0]^2\right)\right\}$	(9)

* CI—competitive inhibition; NCI—noncompetitive inhibition; UCI—uncompetitive inhibition; MI—mixed inhibition.

Only data points in the "linear" part of each curve [P] vs. time were used for subsequent kinetic analysis [10,12].

2.3. Data Processing and Analysis

Based on the same set of simulated values, obtained as explained in Section 2.2, conventional kinetics based on the initial velocities (MME) was performed similarly to a previously published work [13]. The same data points were also submitted to a kinetic analysis using integrated equations (IMME) [8,10], assuming the presence of two inhibitors (mutually exclusive when one of which is a reaction product) and considering the following models: MI model (Equation (9)); UCI (Equation (8)); NCI (Equation (7)); CI (Equation (6)); and WI (Equation (1)).

The estimation of kinetic constants was carried out through the nonlinear regression of the experimental data using the Solver supplement of Microsoft Office Excel [13]. The discrimination between different kinetic models was performed using the Akaike information criterion (AIC) [8,14], as previously explained. Kinetic parameter uncertainties were determined using the sequential quadratic programming algorithm from SPSS Statistics 23 (IBM, Armonk, NY, USA) software.

3. Results

3.1. Simulated Data for the Graphical Representation of Product Versus Time

Assuming the kinetic inhibition by the reaction product (Figure 1a) or considering the simultaneous presence of two mutually exclusive inhibitors when one of which is a reaction product (Figure 1b), this work intends to highlight possible inaccuracies when kinetic studies are carried out using conventional methodologies (initial velocities) [13]. The integrated equations with one inhibitor (Figure 1a) are presented in Table 1. Model MI (Equation (5)) can be simplified into other nested models with fewer parameters (Table 1), assuming that constants K_{ic} and K_{iu} both tend to infinity (WI, Equation (1)), that only one of them does (CI, $K_{iu} \rightarrow \infty$, Equation (2) or UCI, $K_{ic} \rightarrow \infty$, Equation (4)), or that both are equal (NCI, $K_{ic} = K_{iu}$, Equation (3)).

To clarify the need for a new approach to the study of enzyme kinetics when reaction product inhibition occurs, a simulation (Figure 2) was carried out using integrated equations (Table 1) considering the absence of product inhibition (Equation (1)) and admitting, e.g., that the reaction product is a competitive inhibitor (Equation (2)).

3.2. Kinetic Studies Using MME and IMME Based on Simulated Values

When reactions are carried out in the presence of a second inhibitor, another set of integrated equations assumes the simultaneous presence of two mutually exclusive inhibitors (Figure 1b), where one of which is a reaction product. This process is presented in Table 2 [8]. In the presence of two inhibitors, a linear MI model (Equation (9), Table 2) can be simplified into other nested models with fewer parameters (Equations (1), (6)–(8)), assuming that constants K_{ic1}, K_{ic2}, K_{iu1}, K_{iu2} tend to infinity (WI, Equation (1)) or only some of them do (CI or UCI, (Equations (6) and (8)), or when $K_{ic1} = K_{iu1}$ and $K_{ic2} = K_{iu2}$ (NCI, Equation (7)).

The initial velocities obtained with the theoretically simulated data points without or with a second inhibitor were used for the kinetic investigation using a conventional MME. Discrimination among the different kinetic models (Table 3,) using AIC methodology can be seen in Table 4. Uncompetitive inhibition (MME) was discriminated as the best fit model and the estimated parameters are presented in Table 5. The same discrimination methodology for the different IMME models (Table 3) shows that the best model is the MI model (Equation (9)). In addition to discriminating between different models (UCI and MI, respectively, by the MME and IMME methodologies), the results show notable differences between the values of the kinetic parameters determined for both models (Table 5).

Table 3. Sum of square error (SSE) values for different models obtained by the MME and IMME methodologies.

	WI	CI	NCI	UCI	MI
			MME		
SSE	0.001599	0.000340	0.000033	0.000010	0.000007
p	2	3	3	3	4
n	14	14	14	14	14
			IMME		
SSE	946.511	339.962	19.814	3.064	0.0000006
p	2	4	4	4	6
n	432	432	432	432	432

Table 4. Discrimination of the UCI model using AIC methodology with initial velocities data (MME).

Models A/B	SSE_A	SSE_B	n	$P_A + 1$	$P_B + 1$	AIC_A	AIC_B	$AICc_A$	$AICc_B$	Δ	Probability B Correct
WI/UCI	0.001599	0.000010	14	3	4	−121.1	−190.2	−118.7	−185.7	−67.0	1.00
UCI/MI	0.000010	0.000007	14	4	5	−190.2	−192.4	−185.7	−184.9	0.8	0.397

Table 5. Summary of obtained constants with discriminated best fit models.

Method (Model)	K_m (μM)	K_{ic1} (μM)	K_{iu1} (μM)	K_{ic2} (μM)	K_{iu2} (μM)	V_{max} (μmol min^{-1} mg^{-1})
MME (UCI)	344.4 (±17.3)			-	7.3 (±0.3)	0.077 (±0.001)
IMME (MI)	499.99 (±0.001)	99.98 (±0.001)	50.0 (±0.000)	50.0 (±0.000)	5.0 (±0.000)	0.121 (±0.000)

4. Discussion

As can be seen in Figure 2, the experimental points for product versus time cannot be the same once the equations are different. The ratios for K_m/K_{ic} were 1000, 5, 5 and 0.4, respectively to Figure 2a–d. In Figure 2a,b the enzyme is not in saturated conditions, unlike what happens in Figure 2c,d. The kinetic constant K_{ic} (product inhibitor) used to obtain the different curves for Figure 2a is much smaller than K_m. Even when K_m and K_{ic} have the same order of magnitude (Figure 2b,c) or K_{ic} is greater than K_m (Figure 2d), the initial velocities with and without the inhibitor seem similar, but are not the same, as previously discussed [10,12]. In traditional enzyme kinetics lab work, experimental data is usually forced to be adjusted to the Michaelis–Menten equation, i.e., uninhibited conditions are assumed. However, as Figure 2 shows, the initial velocities determined from the derivatives (tangents) of Equation (1) (solid line, without inhibition by the reaction product) and Equation (2) (dashed line, competitive inhibition by the reaction product) at time zero are different mainly when the K_m/K_{ic} ratio is greater than 1 (Figure 2a–c). Therefore, if product inhibition occurs, the initial velocity can never be an experimental point of the Michaelis–Menten equation ($v_0 = V_{max}$ [S]/(K_m + [S]) as assumed by the initial velocities methodology. Unfortunately, all published kinetic studies that use the initial velocities methodology when the product is an inhibitor have this evident incompatibility.

Thus, initial velocity studies in the presence of reaction product inhibition have two implicit drawbacks: (1) the initial velocities of these two equations (without product inhibition and with product inhibition) are the same; and (2) it is assumed that the initial velocities are data points belonging to the Michaelis–Menten equation ($v_0 = V_{max} [S]/(K_m + [S])$). These methodological simplifications were overlooked about 100 years ago when it was very difficult to overcome these problems due to the lack of computers. Even after computer use became common, the determination of initial velocities when the product of the reaction is an inhibitor remains very popular.

In kinetic studies, equations must be adjusted to the data points prior to the discrimination of the best model. Unfortunately, in almost all published studies, data points were adjusted to the equations that were previously assumed to be correct. Thus, miscalculation begins with the initial velocity determinations when the reaction product inhibition is not considered [15,16]. The error can even be accentuated when, in many cases, the data points are adjusted to a particular inhibition model without considering other types of inhibition, which may provide better adjustments (e.g., non-competitive versus mixed inhibition). Inaccuracies are also amplified when the kinetics are studied using archaic curve linearization, such as with the popular Lineweaver–Burk plot, which is still used today [17], or other similar processes. The advantages of integrated equations have been presented by many authors and summarized in previous publications [8–10,12,18–20]. To emphasize the importance of this subject, misinterpretations can be increased when another different inhibitor is added to the reaction medium, which is a mandatory condition in kinetic inhibition studies [8,9]. A summary of the work carried out can be seen in the scheme presented in Figure 3.

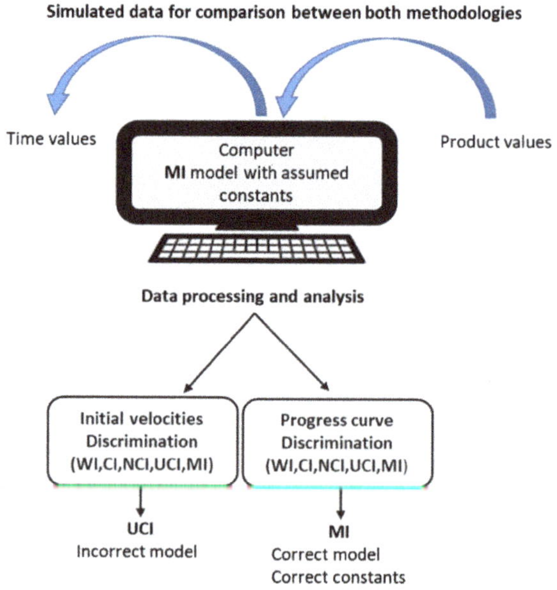

Figure 3. Scheme of theoretical framework developed in order to show that with the same data range, different kinetic models were obtained as a function of the chosen methodology (initial velocities or progress curve analysis).

To exemplify the errors obtained when the initial velocity is assumed, theoretical values (same data points, product versus time) were obtained with Equation (9) (Table 2) and the arbitrary values of the product (P) introduced in the equation (assuming predefined kinetics parameters) in order to obtain time (t) values ($K_m = 500$ µM; $V_{max} = 0.120$ µmol min^{-1} mg^{-1}; $K_{ic1} = 100$ µM; $K_{iu1} = 50$ µM; $K_{ic2} = 50$ µM; $K_{iu2} = 5$ µM). It should be pointed out that ratios

$K_m/K_{ic1} = 5$ and $K_m/K_{iu1} = 10$ are in the same order of magnitude of simulations presented in Figure 2b,c. Moreover, these ratios are in the same order of magnitude exhibited by enzymes with product inhibition [10,11,21]. With these theoretical data points (product vs. time), the study was carried out using the initial velocity methodology (MME) and compared with an IMME based on the same data range (linear phase of product vs. time), i.e., using the same simulated points. The best methodology (MME or IMME) will be the one that provides similar kinetic parameters with those utilised for initial data point generation. As can be seen, the different values for the kinetic parameters and the inhibition model thus discriminated (UCI/MME) do not match the initial constants used to obtain the values of the data set (product vs. time). The kinetic constants obtained by the MME methodology were: $K_m = 344.4$ µM instead of 500 µM; $V_{max} = 0.077$ µmol min^{-1} mg^{-1} instead of 0.120 µmol min^{-1} mg^{-1}; $K_{ic2} = \infty$ instead of 50 µM; and $K_{iu2} = 7.3$ µM instead of 5 µM. On the contrary, the IMME methodology allowed us to obtain identical kinetic parameters (Table 5) and a kinetic model (MI) chosen to generate simulated data points (product vs. time). It is evident that the MME methodology (initial velocity) cannot be used to obtain K_{ic1} and K_{iu1}. In addition, this methodology also failed to discriminate the inhibition model for the second inhibitor (UCI instead of MI) and gave the kinetic parameters (Table 5) K_m and V_{max} with values lower than around 30%. These results highlight the superiority of the IMME methodology.

5. Conclusions

The kinetic analysis of enzymatic reactions in the presence of product inhibition gives different kinetic models as a function of the chosen methodology (initial velocities and progress curve analysis). The results obtained with the MME methodology (initial velocities) gives an inhibition model and parameter constants that do not match with the correct one. In contrast, the IMME methodology allowed us to obtain the same model (MI) and constants similar to the original ones used to obtain the initial data points.

Theoretical considerations regarding the use of integrated Michaelis–Menten equations to study enzyme kinetics when the reaction product is an enzyme inhibitor emphasize the superiority of the IMME approach over the classical MME methodology. Thus, kinetic studies, when the reaction product is an enzyme inhibitor, should be carried out using the IMME methodology in order to obtain the appropriate kinetic model and unbiased kinetic constants.

Author Contributions: Conceptualization, R.M.F.B. and A.A.D.; methodology, R.M.F.B. and A.A.D.; formal analysis, J.M.C.F.; investigation, J.M.C.F.; resources, R.M.F.B. and A.A.D.; writing—original draft preparation, J.M.C.F.; writing—review and editing, J.M.C.F., R.M.F.B. and A.A.D.; supervision, R.M.F.B. and A.A.D. All authors have read and agreed to the published version of the manuscript.

Funding: This work was supported by the National Funds of the FCT—Portuguese Foundation for Science and Technology, under the project UIDB/04033/2020.

Data Availability Statement: Data are available upon request to the authors.

Conflicts of Interest: The authors declare no conflict of interest.

References

1. Adrio, J.L.; Demain, A.L. Microbial enzymes: Tools for biotechnological processes. *Biomolecules* **2014**, *4*, 117–139. [CrossRef]
2. El Harrad, L.; Bourais, I.; Mohammadi, H.; Amine, A. Recent advances in electrochemical biosensors based on enzyme inhibition for clinical and pharmaceutical applications. *Sensors* **2018**, *18*, 164. [CrossRef]
3. Henri, V. Theorie generale de l'action de quelques diástases. *Comptes Rendus Biol.* **1902**, *135*, 916–919.
4. Michaelis, L.; Menten, M. Die kinetik der invertinwirkung. *Biochem. Z.* **1913**, *49*, 333–369.
5. Alam, M.T.; Olin-Sandoval, V.; Stincone, A.; Keller, M.A.; Zelezniak, A.; Luisi, B.F.; Ralser, M. The self-inhibitory nature of metabolic networks and its alleviation through compartmentalization. *Nat. Commun.* **2017**, *8*, 16018. [CrossRef] [PubMed]
6. Andrić, P.; Meyer, A.S.; Jensen, P.A.; Dam-Johansen, K. Reactor design for minimizing product inhibition during enzymatic lignocellulose hydrolysis: II. Quantification of inhibition and suitability of membrane reactors. *Biotechnol. Adv.* **2010**, *28*, 407–425. [CrossRef] [PubMed]

7. Schmidt, N.D.; Peschon, J.J.; Segel, I.H. Kinetics of enzymes subject to very strong product inhibition: Analysis using simplified integrated rate equations and average velocities. *J. Theor. Biol.* **1983**, *100*, 597–611. [CrossRef]
8. Bezerra, R.M.; Pinto, P.A.; Fraga, I.; Dias, A.A. Enzyme inhibition studies by integrated Michaelis–Menten equation considering simultaneous presence of two inhibitors when one of them is a reaction product. *Comput. Meth. Prog. Biomed.* **2016**, *125*, 2–7. [CrossRef]
9. Bezerra, R.M.F.; Pinto, P.A.; Dias, A.A. A kinetic process to determine the interaction type between two compounds, one of which is a reaction product, using alkaline phosphatase inhibition as a case study. *Appl. Biochem. Biotechnol.* **2020**, *191*, 657–665. [CrossRef]
10. Bezerra, R.M.; Fraga, I.; Dias, A.A. Utilization of integrated Michaelis-Menten equations for enzyme inhibition diagnosis and determination of kinetic constants using Solver supplement of Microsoft Office Excel. *Comput. Meth. Prog. Biomed.* **2013**, *109*, 26–31. [CrossRef] [PubMed]
11. Bezerra, R.M.; Dias, A.A.; Fraga, I.; Pereira, A.N. Simultaneous ethanol and cellobiose inhibition of cellulose hydrolysis studied with integrated equations assuming constant or variable substrate concentration. *Appl. Biochem. Biotechnol.* **2006**, *134*, 27–38. [CrossRef]
12. Bezerra, R.M.; Dias, A.A. Utilization of integrated Michaelis-Menten equation to determine kinetic constants. *Biochem. Mol. Biol. Educ.* **2007**, *35*, 145–150. [CrossRef]
13. Dias, A.A.; Pinto, P.A.; Fraga, I.; Bezerra, R.M. Diagnosis of enzyme inhibition using Excel Solver: A combined dry and wet laboratory exercise. *J. Chem. Educ.* **2014**, *91*, 1017–1021. [CrossRef]
14. Pinto, P.A.; Bezerra, R.M.F.; Dias, A.A. Discrimination between rival laccase inhibition models from data sets with one inhibitor concentration using a penalized likelihood analysis and Akaike weights. *Biocatal. Biotransform.* **2018**, *36*, 401–407. [CrossRef]
15. Cao, W.; de La Cruz, E. Quantitative full time course analysis of nonlinear enzyme cycling kinetics. *Sci. Rep.* **2013**, *3*, 2658. [CrossRef]
16. Pinto, M.F.; Ripoll-Rozada, J.; Ramos, H.; Watson, E.E.; Franck, C.; Payne, R.J.; Saraiva, L.; Pereira, P.J.B.; Pastore, A.; Rocha, F.; et al. A simple linearization method unveils hidden enzymatic assay interferences. *Biophys. Chem.* **2019**, *252*, 106193. [CrossRef]
17. Pamidipati, S.; Ahmed, A. A first report on competitive inhibition of laccase enzyme by lignin degradation intermediates. *Folia Microbiol.* **2020**, *65*, 431–437. [CrossRef]
18. Bezerra, R.M.F.; Dias, A.A.; Fraga, I.; Pereira, A.N. Cellulose hydrolysis by cellobiohydrolase Cel7A shows mixedhyperbolic product inhibition. *Appl. Biochem. Biotechnol.* **2011**, *165*, 178–189. [CrossRef] [PubMed]
19. Bezerra, R.M.F.; Dias, A.A. Discrimination among eight modified Michaelis-Menten kinetics models of cellulose hydrolysis with a large range of substrate/enzyme ratios: Inhibition by cellobiose. *Appl. Biochem. Biotechnol.* **2004**, *112*, 173–184. [CrossRef]
20. Bezerra, R.M.F.; Dias, A.A. Enzymatic kinetic of cellulose hydrolysis, Inhibition by ethanol and cellobiose. *Appl. Biochem. Biotechnol.* **2005**, *126*, 49–59. [CrossRef] [PubMed]
21. Prosser, G.A.; de Carvalho, L.P.S. Kinetic mechanism and inhibition of *Mycobacterium tuberculosis* D-alanine: D-alanine ligase by the antibiotic D-cycloserine. *FEBS J.* **2013**, *280*, 1150–1166. [CrossRef] [PubMed]

Article

Compatibility and Washing Performance of Compound Protease Detergent

Wei Zhang [1], Jintao Wu [1], Jing Xiao [1,2,*], Mingyao Zhu [1] and Haichuan Yang [1]

[1] State Key Laboratory of Biobased Material and Green Papermaking, Qilu University of Technology (Shandong Academy of Sciences), Jinan 250353, China; zhangweimickey@163.com (W.Z.); wujintao6666@163.com (J.W.); zhumingyao1999@163.com (M.Z.); 17864114380@163.com (H.Y.)

[2] Shandong Lonct Enzymes Co., Ltd., Linyi 276000, China

* Correspondence: xiaojing8168@163.com; Tel.: +86-0531-8963-1776

Abstract: Protease is the main enzyme of detergent. Through the combination of different proteases and the combination of protease and detergent additives, it can adapt to different washing conditions to improve the washing effect. In this experiment, whiteness determination, microscope scanning, Fourier transform infrared spectroscopy, and X-ray photoelectron spectroscopy were used to detect the whiteness values of the cloth pieces before and after washing, as well as the stain residue between the fibers on the surface of the cloth pieces. The protease detergent formula with better decontamination and anti-deposition effects was selected. The combination of alkaline protease, keratinase, and trypsin was cost-effective in removing stains. Polyacrylamide gel electrophoresis showed that the molecular weight of the protein significantly changed after adding the enzyme preparation during washing, and the molecular weight of the protein was directly proportional to protein redeposition. The composite protease had a better comprehensive decontamination effect, and when compatible with suitable surfactants, anti-redeposition agents, and water-softening agents, the compound protease detergent exhibited a stronger decontamination ability than commercial detergents.

Keywords: protease; detergent; surfactant; cleaning

Citation: Zhang, W.; Wu, J.; Xiao, J.; Zhu, M.; Yang, H. Compatibility and Washing Performance of Compound Protease Detergent. *Appl. Sci.* **2022**, *12*, 150. https://doi.org/10.3390/app12010150

Academic Editor: In Jung Kim

Received: 29 October 2021
Accepted: 18 December 2021
Published: 24 December 2021

Publisher's Note: MDPI stays neutral with regard to jurisdictional claims in published maps and institutional affiliations.

1. Introduction

Protease is added to detergents to decompose stains during laundry [1,2]. The protease breaks the polypeptide chain, the macromolecular protein is broken down into small molecule polypeptides or amino acids, and it is peeled from the fabric under the action of surfactant and external force [3]. The earliest protease used for washing was trypsin, which only needs a small amount to achieve good washing results. However, trypsin is mainly extracted from animal materials [4,5], and the raw materials and processes have certain limitations. Thus, it has been gradually replaced with alkaline protease, which can be produced by large-scale fermentation. Current enzyme-added detergents only add alkaline protease and exhibit a single washing effect. With the large-scale production of multiple varieties of proteases, the production cost of proteases has been reduced. With the improvement of living conditions, people's requirements for washing have increased, and new types of protein stain washing solutions are needed.

In 1963, Novozymes introduced protease, which led to a revolution in the industrial enzyme market and began a rapid expansion of detergent enzyme products. Currently, enzymes for detergents already account for 40% of all industrial enzymes. In the European and American markets, enzyme detergents already account for 80% of the detergent market, and almost all detergents are enzyme detergents in Japan. The current trend of enzymatic detergent research and development is extended from single protease to multiple enzymes, such as lipase, amylase, cellulase, mannanase, peroxidase, laccase, etc., and from a single type of stain to a comprehensive washing for multiple stains [6]. Improving the storage stability of proteases in detergents and washing stability under high/low temperature and high alkaline/acidic conditions have also become new research hotspots. In order to meet the diverse functional requirements of the consumer market for detergents, Novozymes has developed a hybrid liquid detergent Medley solution. The enzymes used in the new Medley solution maintain stability and washing performance at high water content. Alkaline proteases have become key ingredients in detergent formulations [7]. With the emergence of new strains, researchers have found that other proteases can be added to detergents, such as keratinase [8]. The current problems of protease detergents are the compatibility of protease with washing auxiliaries and the stability of protease at different temperatures and pH [9,10].

The effect of trypsin [11], keratinase [12,13], and alkaline protease [14] on cleaning is obvious. The compound washing effect is better after the protease is proportioned. Detergents are added with different ingredients [15], such as surfactants [16], water-softening agents [17], anti-redeposition agents [18], softening agents [19,20], and stabilizers, to improve the washing efficiency. We selected 30 common detergent auxiliaries, matched them with protease, and compared the washing effects of different compound protease detergents, and the optimal formula was obtained. The washing power of the composite protease detergent was better than those of several common commercial detergents [21].

2. Materials and Methods

2.1. Materials

Selected proteases included alkaline protease (5.0×10^5 U/g), weakly alkaline protease (4.8×10^5 U/g), neutral protease (1.5×10^5 U/g), acid protease (8.0×10^5 U/g), trypsin (4.0×10^3 U/g), aminopeptidase (6.0×10^4 U/g), flavourzyme (5.0×10^5 U/g), and keratinase (1.0×10^5 U/g) in food-grade solid powder form from Lonct Enzymes Co., Ltd. (Linyi, China). Protease activity is expressed as protease activity units. At 40 °C and a certain pH, the amount of enzyme required for protease to hydrolyze casein to produce 1 μg of tyrosine per minute is one unit of enzyme activity. The pH of the reaction conditions for measuring the enzymatic activity of alkaline proteases, weakly basic proteases, aminopeptidases, and keratinases was 10.5. The pH of the reaction conditions for measuring the enzymatic activity of neutral proteases, trypsin, and flavored proteases was 7.5. The pH of the reaction conditions for measuring the enzymatic activity of acidic proteases was 3.0.

Detergent builders included polyethoxylated fatty alcohols, sodium ethoxy alkyl sulfate, dodecylbenzenesulfonic acid, triethanolamine, anhydrous sodium citrate, SNS-80, each in industrial-grade ($\geq$90%) from Research Institute of Daily Chemical Industry (Taiyuan, China). 2-morpholineethanesulfonic acid (MES) in molecular biology grade (>99%) from Coolaber. Fatty acid methyl ester ethoxylate (FMEE), alkyl glycoside (APG), layered sodium disilicate (SKS-6) from Yousuo Chemical Technology Co., Ltd. (Linyi, China), modified oil ethoxylate (SOE) from Junxin Chemical Technology Co., Ltd. (Guangzhou, China), tea saponin from Zhongye Biotechnology Co., Ltd. (Lishui, China), sodium alginate, hyaluronic acid from Boxbio Science & Technology Co., Ltd. (Beijing, China), all of the above are industrial grade ($\geq$90%). Sodium carboxymethyl cellulose (CMC), polyvinyl alcohol type 1799 (PVA), polyethylene glycol 6000 (PEG), polyaspartic acid (PASP), silicon dioxide (SiO_2), ethylenediaminetetraacetic acid tetrasodium salt (EDTA), disodium maleate were used in chemically pure forms ($\geq$99.5%) from Sinopharm Chemical Reagent Co., Ltd. (Shanghai, China). Hydroxypropyl methylcellulose sodium (HPMC), hydroxyethyl cellulose sodium, and sodium polyacrylate were used in food-grade (>99%) from Best Food Additives Co., Ltd. (Zhengzhou, China). 4A zeolite was industrial-grade ($\geq$90%) from Runfeng Synthetic Technology Co., Ltd. (Nantong, China). Sodium tartrate and sodium gluconate in food-grade (>99%) were acquired from Gukang Biological Engineering Co., Ltd. (Jinan, China). Sodium laurate in industrial-grade ($\geq$90%) was from Longhui Chemical Co., Ltd. (Jinan, China). α-cyclodextrin, β-cyclodextrin, and γ-cyclodextrin were used in food-grade (>99%) from Youlezi Food Ingredients Co., Ltd. (Shanghai, China). Sulfobutyl-β-cyclodextrin (Captisol), methyl-β-cyclodextrin, and 2-hydroxypropyl-β-cyclodextrin were analytical reagents ($\geq$99.9%) from Aladdin (Shanghai, China). In the experiment, the materials used to simulate protein stains were eggs and carbon powder. Blood used was porcine anticoagulated whole blood. Double distilled water was used in the experiments.

2.2. Preparation of Soiled Cloth

The protein stain was prepared as follows: Weigh 2.4 g of gum arabic powder and dissolve it with a little water, add 1.6 g of carbon black powder and grind for about 2 min. Transfer this carbon black stain to 120 mL of aqueous solution containing 13.8 g of whole milk powder, add another 120 mL of distilled water, homogenize with an emulsifier at 4000–5000 r/min for 30 min, then slowly add 120 mL of aqueous solution containing 25 g of egg liquid (egg white: yolk = 3:2) and continue to homogenize for 1 h.

White cotton cloth was cut into circular pieces of ~6 cm. When preparing the protein fouling cloth, the protein stain solution was heated to 40 °C and filtered. Then, 200 μL was added dropwise onto the white cotton cloth, soaked, pressed, and then dried. The same method was followed when preparing the dirty blood cloth.

2.3. Washing Procedure

The preparation method of the basal detergent is as follows. Add 4% polyethoxylated fatty alcohol, 2% ethoxylated alkyl sulfate, 8% dodecylbenzene sulfonic acid, 0.5% triethanolamine, and 0.5% anhydrous sodium citrate in a volume of water, stir to dissolve, and use sodium hydroxide solution to adjust the pH of the solution to 8.5–9.0.

The water used during washing was hard water (250 mg/kg), and the molar ratio of Ca^{2+} to Mg^{2+} was 6:4. The configuration method is as follows: weigh 1.67 g $CaCl_2$ and 2.04 g $MgCl_2 \cdot 6H_2O$, add water to make 10.0 L, which is 250 mg/kg hard water.

Then, 400 U/g of protease was added to the basal detergent and mixed well to prepare a 0.2% solution. A 100 mL aliquot of the solution was added to an Erlenmeyer flask, and a cloth piece was added. Various proteases have their maximum enzyme activity at 50–60 °C. Thus, 50 °C was selected as the reaction temperature [22,23]. The rotating speed was maintained at 150 r/min at 50 °C and washed for 50 min. Each piece of cloth was rinsed, dehydrated, and then dried.

2.4. Characterization of the Cleaning Effect

A whiteness tester was used to detect the whiteness reflectance value of the cloth surface before and after washing at the wavelength of 457 nm. We took two points on the front and back sides of the cloth piece before washing or after washing and measured the whiteness value. The average value of the four measurements is the whiteness value of the cloth piece. The stain residues on the surface and inside of the fiber before and after washing were observed with a super depth-of-field microscope (Leica DVM6A, Chongqing, China) at magnifications of 200× and 500× [24]. The state of the fibers before and after washing the cloth and the stain residues between the fibers were observed via scanning electron microscopy (SEM, Phenom pure plus, Shanghai, China) at a voltage of 10 kV [13,25]. The magnifications were 410×, 430×, and 440×. Fourier transform infrared spectroscopy (ATR-FTIR, Nicolet10, Waltham, MA, USA) was performed before and after washing of the fabric sheet [26,27]. The wavenumber range of residual functional groups was 500–4000 cm^{-1}. Energy dispersion analysis was performed using X-ray photoelectron spectrometry (XPS, ESCALABXi+, Waltham, MA, USA) [27].

2.5. Detection of Stain Protein Molecular Weight

Eighty units of enzyme activities of alkaline protease, keratinase, and 4 U enzyme activities of trypsin were added to the protein-contaminated liquid to verify the decomposing effect of the proteases on protein stains. The reaction was carried out at 50 °C for 50 min, and the reaction solution was analyzed by SDS–PAGE [14,28]. Eighty units of alkaline protease, keratinase, and trypsin were added to the blood to verify the decomposing effect of the proteases on bloodstains. The reaction was carried out at 50 °C for 10–50 min, and the reaction solution was analyzed by SDS–PAGE. The electrophoresis gel was prepared using the Meilun protein gel kit. The sample was added with 80 μL Tricine-SDS-PAGE loading buffer (5×), boiled for 5 min, centrifuged to obtain the supernatant, and then loaded with 10–15 μL for each sample. Electrophoresis was performed after adding the electrophoresis buffer to the electrophoresis system.

2.6. Evaluation of the Effect of Blood Stains Redeposition

5 mL of 4% blood dilution solution was prepared; added with 80 U of alkaline protease, keratinase, and trypsin; and then shaken in a water bath at 50 °C for 10, 20, 30, 40, and 50 min. The reaction solution was boiled for 3 min, and then protein gel electrophoresis was performed. Double-distilled water (5 mL) was added to the reaction solution, in which the white cotton cloth was immersed and allowed to stand for 12 h at 30 °C. The cloth was rinsed with water and dried, and then the extent of surface deposition and the relationship between the extent of protein deposition on the surface of the cloth and the molecular weight of the protein were observed.

2.7. Optimization of Protease Washing Performance

Alkaline protease and keratinase were mixed in the detergent at different ratios, and the total enzyme activity of each experiment was set to 80 U. In the experiment, the enzyme activity ratios of alkaline protease and keratinase were 20 U:60 U, 40 U:40 U, 30 U:50 U, 60 U:20 U, and 10 U:70 U. On the basis of adding 80 U protease activity per 0.2 g detergent, the enzyme activity of trypsin was added as 5 U, 10 U, 15 U, 20 U successively, and the remaining enzyme activity was supplemented to 80 U by alkaline protease and keratinase. The selected detergent auxiliaries were added to the basal detergent, and the addition amount was 1% of the mass of basal detergent. One or two additives, enzymes, and basal detergents with the best cleaning effect were selected from the surfactants, anti-deposition agents, water softeners, and cyclodextrins for subsequent experiments. The optimized formula was compared with four common commercial detergents in terms of washing effect.

3. Results and Discussion

3.1. Washing Performance Test of Different Proteases

The washing effect of different proteases on protein fouling cloths and dirty blood cloths was analyzed. Saleem et al. [29] reported the capability of protease to digest and convert the insoluble form of egg white and blood clot into their soluble forms. Bersic et al. [30] reported that adding protease to detergent has a better washing effect. Through the above washing method, different proteases were used to clean protein-fouling cloths and blood-dirty cloths under the conditions of 80 U, 50 °C, and 50 min (Table 1). In the washing of protein stains, the highest reflectivity was obtained when trypsin was added, equivalent to 1.3 times upon the addition of alkaline protease, followed by keratinase. In the washing of bloodstains, the highest reflectivity was obtained when trypsin was added, equivalent to 1.6 times upon the addition of alkaline protease, followed by keratinase. Neutral protease and aminopeptidase have the worst washing effects. In washing blood-stained cloths with protease, some proteases could not completely decompose hemoglobin on the cloth piece, which caused hemoglobin to be retained by the fibers on the surface of the cloth piece and redeposited on the surface of the cotton cloth. Paul et al. [13] reported that crude keratinase could effectively remove blood and egg yolk stains and can be added to detergent products as a washing aid. In addition, the sewage used for washing does not pollute water resources. Emran et al. [31] reported that alkaline protease is the most commonly used protease in detergents because of its good thermal stability and compatibility with detergents. However, continuing research and development on detergents is important to find proteases with better effects. Commercial keratinase and trypsin have become promising choices. As a protease added to detergents, commercial keratinase is a more promising option. Trypsin has better detergent effects but is limited by price and production conditions and can be added to special detergents as appropriate.

Table 1. Washing performance of protease on protein-fouling and blood-dirty cloths.

Washing Performance	Reflectivity (%)	
	Protein-Fouling Cloth	Blood-Dirty Cloth
Before washing	13.23 ± 1.12	10.33 ± 1.0
Basal detergent	20.90 ± 0.56	26.20 ± 1.76
Alkaline protease	25.60 ± 0.59	25.20 ± 1.39
Keratinase	29.38 ± 0.79	29.58 ± 2.50
Trypsin	32.90 ± 1.90	41.08 ± 0.60
Weak alkaline protease	21.95 ± 0.75	18.43 ± 1.72
Neutral protease	19.10 ± 0.91	21.45 ± 2.10
Acid protease	20.85 ± 0.96	22.48 ± 2.12
Aminopeptidase	20.83 ± 0.51	19.93 ± 1.98
Flavour protease	23.90 ± 1.53	30.40 ± 0.71

The stain residue on the surface and interior of the protein fouling cloth and blood-dirty cloth before cleaning and after protease washing was observed under an ultra-depth-of-field microscope and a desktop scanning electron microscope (Figure 1). A large number of stains were attached to the fiber surface and between the fibers before washing, and the amount of residual stains on the surface and between the fibers of the experimental group after washing with enzymes was significantly reduced, and the fibers were arranged loosely and smoothly. The cleaning effect of the cloth after washing with trypsin and keratinase was better than that after washing with alkaline protease.

The total reflectance of the protein-fouling cloth and blood-dirty cloth before and after washing with different proteases was analyzed using FTIR in the wavenumber range of 500–4000 cm^{-1}. McCutcheon et al. [26] reported that the special spectral characteristics of protein detected by FTIR can reflect the residual protein stains on the surface of the fabric and the interior of the fabric fiber. ATR-FTIR analysis before washing the dirty cloth showed significant infrared absorption peaks at 1640 and 1526 cm^{-1}, which represent the two infrared absorption peaks in the protein-peptide bond C=O stretching vibration absorption peaks and β-sheet conformation amide III band characteristic absorption band (Figure 2). The area of absorption peaks at 1640 and 1526 cm^{-1} of the stained cloth washed with enzymes was significantly reduced, indicating that protein stains were separated from the cloth after being decomposed by enzymes. Alkaline protease, keratinase, and trypsin can all decompose proteins to different degrees, but trypsin and keratinase were better than alkaline protease in decomposing proteins.

Elemental analysis on the cloth was performed using XPS before and after washing with alkaline protease, keratinase, and trypsin (Figure 3). The protein-fouling cloth and blood-dirty cloth before washing showed obvious peaks at 398–400 eV, indicating that they contain N elements. After washing with protease detergent, the peaks of N elements were significantly reduced. This result indicates that the protein on the surface of the dirty cloth was decomposed and removed by the proteases.

3.2. Evaluation of the Ability of Protease to Resist Stain Redeposition

The reaction solution was analyzed by SDS–PAGE (Figure 4), the molecular weight of alkaline protease was approximately 30 kDa, and the molecular weight of keratinase was approximately less than 15 kDa. After the reaction of protein stain and alkaline protease, the molecular weight of the protein was concentrated in 50–80 kDa, and a small part of the molecular weight was concentrated in 25–30 kDa. After the protein stain liquid was hydrolyzed by keratinase, only a small amount of protein had a molecular weight of 30 kDa. This result indicates that the protein-decomposing effect of keratinase is significantly better than that of alkaline protease. The 4 U enzyme activity of trypsin was reacted with the protein stain solution. The molecular weight of the protein after the reaction was less than 25–35 kDa. The amount of trypsin added was much smaller than those of alkaline protease and keratinase, but the protein stains were decomposed effectively. A comparison of the SDS-PAGE results with the application results showed that the protease with a better washing effect could break down protein stains from large molecules into small molecules, and the stains were easier to remove under the action of detergents.

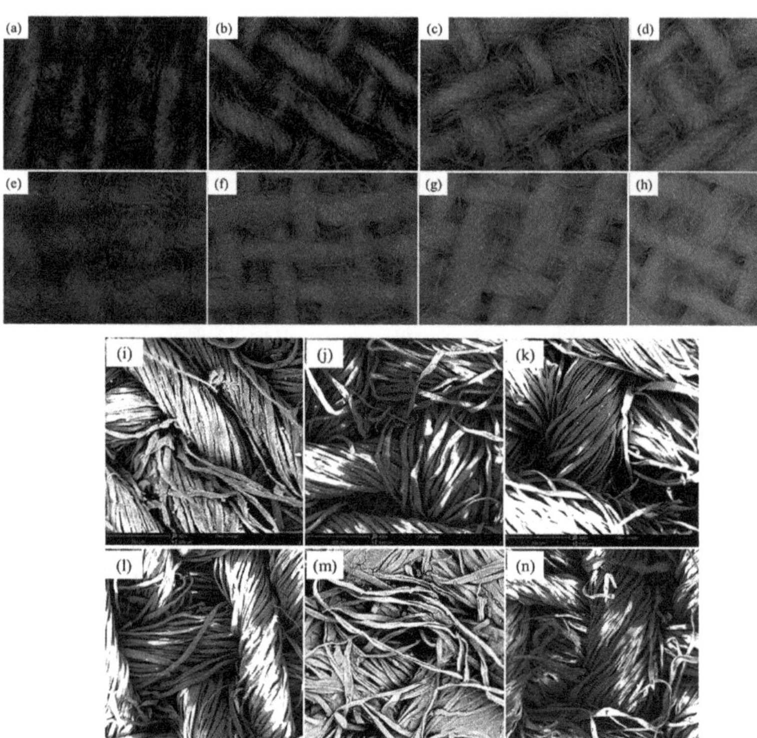

Figure 1. Surface image of protein-fouling cloth. (**a**) Unwashed protein-fouling cloth (500×). (**b**) Protein-fouling cloth after washing with alkaline protease (500×). (**c**) Protein-fouling cloth after washing with keratinase (500×). (**d**) Protein-fouling cloth after washing with trypsin (500×). Surface image of blood-dirty cloth (**e**) Unwashed blood-dirty cloth (500×). (**f**) Blood-dirty cloth after washing with alkaline protease (500×). (**g**) Blood-dirty cloth after washing with keratinase (500×). (**h**) Blood-dirty cloth after washing with trypsin (500×). SEM images of protein-fouling cloth before and after washing. (**i**) Unwashed protein-fouling cloth (420×). (**j**) Protein-fouling cloth after washing with alkaline protease (420×). (**k**) Protein-fouling cloth after washing with keratinase (420×). (**l**) Protein-fouling cloth after washing with trypsin (420×). (**m**) Blood-dirty cloth after washing with keratinase (440×). (**n**) Blood-dirty cloth after washing with trypsin (410×).

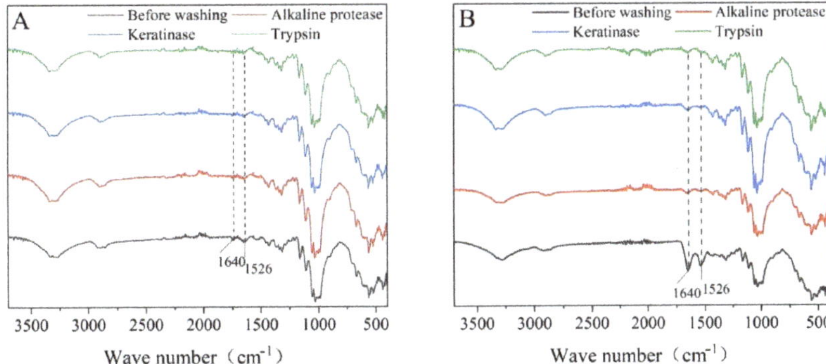

Figure 2. Infrared spectra of dirty cloth before and after washing with different proteases. (**A**) Protein-fouling cloth. (**B**) Blooddirty cloth.

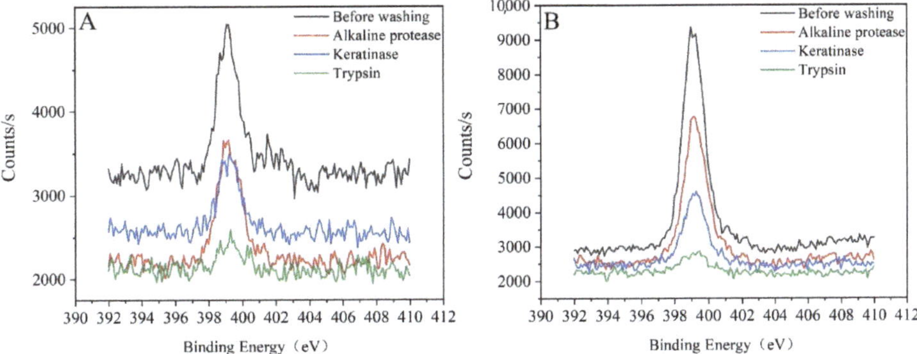

Figure 3. X-ray photoelectron spectroscopy of dirty cloth before and after washing with different proteases. (**A**) Protein-fouling cloth. (**B**) Blood-dirty cloth.

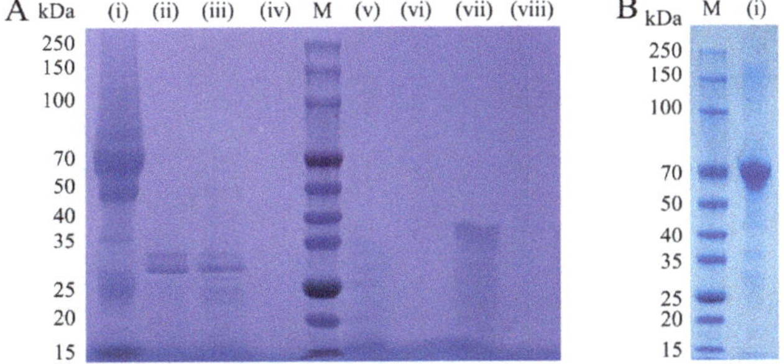

Figure 4. SDS-PAGE result graph. (**A**) M is Marker. (i) Protein stain and alkaline protease. (ii) Alkaline protease and double-distilled water. (iii) Protein stain and keratinase. (iv) Keratinase and double-distilled water. (v) Protein stain and alkaline protease and keratinase. (vi) Alkaline protease and keratinase and double-distilled water. (vii) Protein stain and trypsin. (viii) Trypsin and double-distilled water. (**B**) M is Marker. (i) Protein stain.

The bloodstains on the surface of the cloth treated with alkaline protease, keratinase, and trypsin were compared (Figure 5). Protein molecular weights after treatment with alkaline protease were 70 and 25 kDa. Bloodstains were deposited on the surface of the cloth, but the deposit was relatively reduced as the enzymatic hydrolysis time was prolonged. Bloodstains were analyzed after treatment with keratinase. The protein molecular weights were mainly 50, 40, and 25 kDa. The bloodstains were deposited in the cloth piece. As the enzymatic hydrolysis time was prolonged, the deposition gradually decreased. Compared with the alkaline protease, there were less bloodstains deposited after the keratinase hydrolysis. The solution was treated with trypsin precipitation, seen from the electrophoresis pattern analysis, after 10 min, 20 min, and 30 min reactions. The protein molecular weights were mainly 50 kDa, 22 kDa, 15 kDa, 10 kDa, and macromolecular protein gradually reduced; after 40 min and 50 min reactions, protein molecular weight was mainly concentrated below 10 kDa. Blood from the cloth sheet deposition of view was observed, and the trypsin-treated blood was not deposited onto the fabric sheet again.

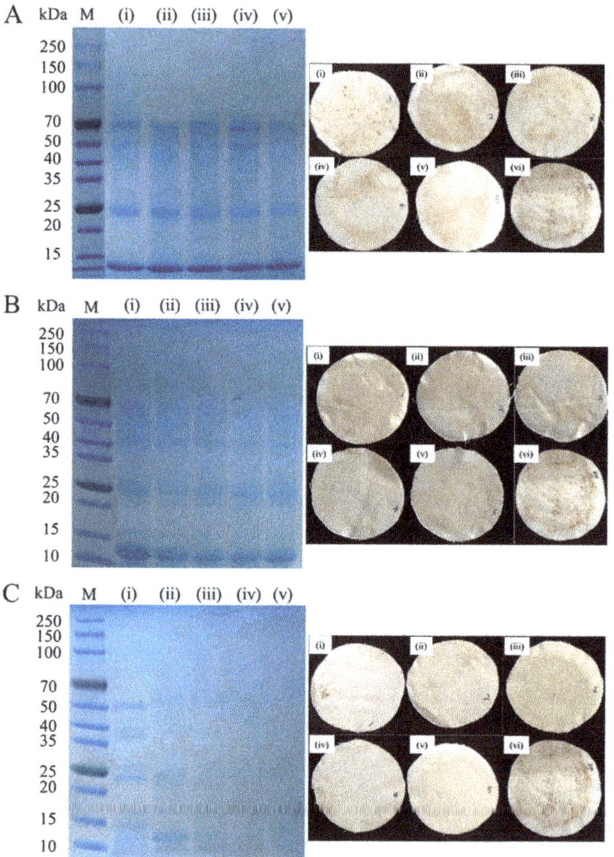

Figure 5. Deposition effect after the blood is decomposed by proteases. (**A**) Alkaline protease. (**B**) Keratinase. (**C**) Trypsin. (i) Reaction 10 min. (ii) Reaction decomposition 20 min. (iii) Reaction 30 min. (iv) Reaction 40 min. (v) Reaction 50 min. (vi) No added protease.

3.3. Alkaline Protease and Detergent Auxiliary Washing Effect Test

The effect of detergents mixed with different detergent auxiliaries and alkaline protease ratios to clean protein-soiled cloth and blood-dirty cloth was tested (Tables 2 and 3). The evaluation criterion is whether the decontamination effect of the mixed detergent is better than that of only the alkaline protease detergent. Better washing surfactants include sodium alginate, FMEE, SNS-80, hyaluronic acid, APG [32], SOE, MES, and tea saponin. Water-softening agents with a better washing effect are 4A zeolite [33,34], sodium gluconate, sodium tartrate, and sodium laurate. The ones that inhibit the effect of alkaline protease are SKS-6, SiO_2, and EDTA. Experimental results show that in the washing process of using alkaline protease detergent, whether it is for protein or bloodstains, it can be mixed with water-softening agent sodium gluconate, surfactant sodium alginate, FMEE, hyaluronic acid, APG, MES, and tea saponin [35,36]. In addition, bloodstain washing can be mixed with anti-deposition agent HPMC, sodium hydroxyethyl cellulose [37], sodium polyacrylate, and PVA. SOE, SiO_2, CMC, disodium maleate, PASP, PEG, EDTA, and EDTA exert inhibitory effects on the protein decomposing effect of alkaline protease.

Table 2. Washing effect of different surfactants and water-softening agents mixed with alkaline protease on dirty cloth.

Washing Performance		Reflectivity (%)	
		Protein-Fouling Cloth	**Blood-Dirty Cloth**
	Before washing	13.23 ± 1.12	10.78 ± 1.01
	Basal detergent	20.90 ± 0.56	14.93 ± 1.30
	Alkaline protease	25.60 ± 0.59	25.20 ± 1.39
	APG	27.03 ± 0.46	24.50 ± 1.67
	SOE	26.90 ± 1.70	21.98 ± 2.12
	MES	26.90 ± 0.59	25.18 ± 0.86
Surfactant	SNS-80	28.55 ± 1.10	23.98 ± 2.32
	FMEE	29.28 ± 2.00	30.15 ± 1.90
	Sodium Alginate	28.58 ± 1.40	27.00 ± 0.76
	Hyaluronic acid	25.40 ± 1.26	18.35 ± 1.94
	Tea saponin	23.93 ± 1.52	24.48 ± 1.29
	EDTA	18.86 ± 2.01	12.56 ± 1.31
	SiO_2	21.34 ± 0.28	12.64 ± 1.11
	SKS-6	22.43 ± 1.25	14.79 ± 1.01
Water-softening agent	Sodium Tartrate	24.95 ± 0.68	14.64 ± 1.36
	Sodium laurate	24.20 ± 1.39	16.65 ± 1.38
	Sodium Gluconate	25.58 ± 1.01	16.69 ± 1.48
	4A zeolite	26.71 ± 1.57	13.80 ± 1.28

Table 3. Washing effect of anti-redeposition agent mixed with alkaline protease on blood dirty cloth.

Washing Performance		Reflectivity (%)
	Before washing	10.78 ± 1.04
	Basal detergent	14.93 ± 1.36
	Alkaline protease	25.20 ± 1.39
	PASP	11.81 ± 1.13
	PEG	11.43 ± 1.05
	CMC	14.24 ± 1.32
Anti-redeposition agent	PVA	15.06 ± 1.56
	HPMC	24.40 ± 2.04
	Sodium Hydroxyethyl Cellulose	21.18 ± 2.12
	Sodium polyacrylate	19.14 ± 1.67
	Disodium Maleate	13.18 ± 1.21
	α-cyclodextrin	26.04 ± 2.41
	β-cyclodextrin	23.38 ± 2.05
Cyclodextrin	γ-cyclodextrin	22.50 ± 2.08
	Sulfobutyl-β-cyclodextrin	31.44 ± 0.47
	Methyl-β-cyclodextrin	24.76 ± 2.49
	2-hydroxypropyl-β-cyclodextrin	21.51 ± 1.99

Adding cyclodextrin to the detergent can help remove blood stains. A variety of cyclodextrins [38] mixed with alkaline protease can be used to wash blood-dirty cloth. Results showed that the addition of α-cyclodextrin and β-cyclodextrin to alkaline protease detergent was better than that of γ-cyclodextrin. The solubility of α-cyclodextrin and β-cyclodextrin was not good (Table 3), and 2-hydroxypropyl-β-cyclodextrin showed better water solubility than β-cyclodextrin. Adding sulfobutyl-β-cyclodextrin to the alkaline protease detergent demonstrated a better washing effect, but it is currently mainly used in the medical field and is expensive.

3.4. Trypsin, Keratinase, and Detergent Auxiliaries Washing Effect Test

The cleaning effect of a single alkaline protease is limited. The compatibility of detergent auxiliaries suitable for alkaline protease with keratinase and trypsin must be tested to improve protease compounding and compatibility with detergent auxiliaries. The protein-fouling and blood-dirty cloths were washed after mixing the detergent auxiliaries and trypsin (Table 4 (a)). The evaluation criterion is whether the washing effect of the mixed detergent is better than that of the detergent only added with trypsin. Results showed that the washing effect of the mixed detergent was better than that of only trypsin detergent. The effect of washing a blood-dirty cloth with mixed detergent was not much different from that of only adding trypsin detergent. Adding a detergent to trypsin detergent exerted no obvious effect on washing blood-dirty cloths.

Table 4. Washing effect of detergent auxiliaries mixed with trypsin or keratinase on dirty cloths.

Washing Performance		Reflectivity (%)	
		(a) Trypsin	(b) Keratinase
Protein-fouling cloth	Enzyme	32.90 ± 0.51	29.38 ± 1.53
	Before washing	13.23 ± 1.12	13.23 ± 1.12
	Basal detergent	20.90 ± 0.56	20.90 ± 0.56
	Hyaluronic acid	32.45 ± 1.18	36.53 ± 3.55
	Sodium Alginate	37.13 ± 1.80	40.35 ± 1.84
	Sodium Gluconate	35.53 ± 2.01	38.83 ± 1.69
	4A zeolite	32.68 ± 1.51	35.38 ± 1.47
Blood-dirty cloth	Enzyme	37.60 ± 0.88	25.20 ± 2.50
	Before washing	10.78 ± 1.04	10.78 ± 1.06
	Basal detergent	14.93 ± 1.37	14.93 ± 1.34
	Sodium Hydroxyethyl Cellulose	35.85 ± 1.14	29.93 ± 1.37
	Sodium Alginate	35.68 ± 1.60	26.43 ± 2.56
	Hyaluronic acid	35.85 ± 0.17	24.40 ± 2.12
	HPMC	35.78 ± 1.14	21.18 ± 2.10
	Sodium polyacrylate	35.35 ± 0.31	27.75 ± 2.66
	CMC	36.75 ± 1.05	28.68 ± 2.40

The protein-fouling and blood-dirty cloths were washed after mixing the detergent auxiliaries and keratinase (Table 4 (b)). The evaluation standard is whether the cleaning effect of the detergent after mixing is better than that of using only the keratinase detergent. For washing protein stains, it can be mixed with sodium alginate, sodium gluconate, hyaluronic acid, and 4A zeolite. For washing bloodstains, it can be mixed with sodium hydroxyethyl cellulose and CMC, followed by sodium polyacrylate, sodium alginate, hyaluronic acid, and HPMC.

3.5. Compound Protease Washing Test

The washing effect of alkaline protease alone is limited. Niyonzima et al. [39] reported that using multiple enzymes to work together in the washing process can decompose and remove stains efficiently. In this experiment, a compound detergent formula composed of alkaline protease, keratinase, and trypsin was prepared, and the total enzyme activity of each group of experiments was set to 80 U. In the experiment, the mixing ratios of alkaline protease and keratinase were 2:6, 4:4, 3:5, 6:2, and 1:7. Stain removal showed a slight difference when alkaline protease and keratinase were added in different proportions to the detergent during washing protein-soiled cloths. When the detergent alkaline protease was mixed with keratinase at a ratio of 1:7, the effect was greatest in washing soiled cloths with blood (Table 5).

Table 5. Cleaning effect of protease compound on protein-fouling cloth and blood-dirty cloth.

Washing Performance	Reflectivity (%)	
	Protein-Fouling Cloth	**Blood-Dirty Cloth**
Alkaline protease	25.60 ± 0.59	25.20 ± 1.39
Keratinase	29.38 ± 1.53	25.20 ± 2.45
Alkaline Protease: Keratinase		
6:2	30.71 ± 0.67	21.96 ± 2.12
4:4	30.33 ± 0.43	24.99 ± 2.45
3:5	30.57 ± 0.31	29.55 ± 2.80
2:6	30.11 ± 1.81	25.34 ± 2.39
1:7	36.28 ± 1.02	34.03 ± 0.62
(Alkaline protease + keratinase) + trypsin		
75 + 5	35.78 ± 1.37	38.30 ± 1.14
70 + 10	40.38 ± 2.52	39.50 ± 0.71
65 + 15	36.73 ± 1.29	39.23 ± 1.40
60 + 20	34.10 ± 2.26	43.43 ± 0.51

The detergent added with trypsin exerted obvious washing effects on protein stains. The addition amount of trypsin ranged from 5 U to 20 U. Each group to add 80 U enzyme activities of protease, wherein the proportion of alkaline protease and keratinase added in the detergent was set to 1:7. Four experimental groups had 5, 10, 15, and 20 U of trypsin added. When the amount of trypsin added in the compound detergent was 10 U, the effect of washing the protein-soiled cloth improved. When the addition amount of the composite detergent trypsin was 5 U, the washing effect with bloodstains became more pronounced (Table 6).

Table 6. Protease liquid detergent sample composition and ratio (%).

Sample Number	A	B	C	D
Polyethoxylated fatty alcohol	4	4	4	4
Sodium alcohol ether sulfate	2	2	2	2
Sodium dodecyl benzene sulfonate	8	8	8	8
Triethanolamine	0.5	0.5	0.5	0.5
Keratinase	0.35	0.35	0.35	0.35
Alkaline protease	0.01	0.01	0.01	0.01
Sodium Alginate	1	1	1	1
Sodium Gluconate	1	0	1	0
Sodium Carboxymethyl Cellulose	1	1	0	0
Sodium Hydroxyethyl Cellulose	0	0	1	1
4Azeolite	0	1	0	1
α-cyclodextrin	1	1	1	1
Anhydrous sodium citrate	0.5	0.5	0.5	0.5
Water	Margin	Margin	Margin	Margin

3.6. Comparison of the Effect of Detergent Products

Liquid detergents usually have surfactants, enzymes, water-softening agents, and anti-redeposition agents added to improve their effect [40]. The current source of trypsin is mainly extracted from animal tissues, which is relatively expensive. When the ratio of alkaline protease and keratinase was set to 1:7, the washing effect can be close to the effect of using only trypsin. Therefore, alkaline protease and keratinase were selected for the compound experiment. One or two detergent auxiliaries, enzymes, and basal detergents were combined to improve washing effects in the above experiments. Four liquid detergents with the highest sales volume on the e-commerce platform and the mixed protease detergent samples were selected to compare the effects and verify the decontamination performance of the composite detergent. The whiteness value of the protein-fouling cloth washed with formulas A, B, C, and D were significantly higher than the whiteness value of the protein-fouling cloth after washing with the selected four commercial detergents. The whiteness value of the blood-dirty cloths washed with formula B, C, and D were slightly higher than that of samples 1 and 4 and was basically the same as that of samples 2 and 3. The whiteness value of the blood-dirty cloths washed with formula A, B, C, and D were slightly higher than the whiteness value of the -cloths after washing with the selected four commercial detergents. Formula A exhibited stronger washing effects and detergency than commercial detergents (Table 7, Figure 6).

Table 7. Cleaning effect of protease liquid detergent samples and commercial liquid detergents on dirty cloth.

Washing Performance	Reflectivity (%)	
	Protein-Fouling Cloth	**Blood-Dirty Cloth**
A	33.43 ± 0.51	38.08 ± 1.76
B	32.05 ± 2.13	36.38 ± 1.43
C	29.85 ± 1.73	35.83 ± 1.99
D	32.38 ± 1.41	37.55 ± 0.92
Sample 1	20.45 ± 1.63	31.15 ± 1.82
Sample 2	24.90 ± 0.52	35.30 ± 2.77
Sample 3	21.58 ± 0.54	36.03 ± 2.80
Sample 4	22.40 ± 1.34	32.82 ± 3.30

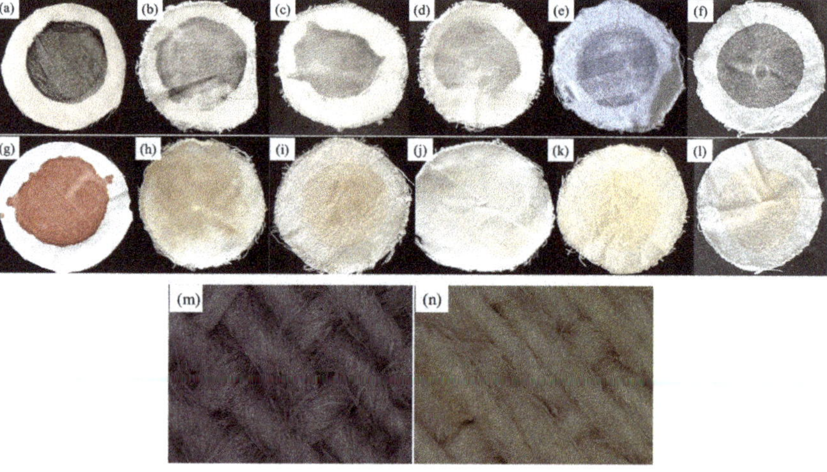

Figure 6. Surface image of protein-fouling cloth. (**a**) Unwashed protein-fouling cloth. (**b**) Protein-fouling cloth after alkaline protease washing. (**c**) Protein-fouling cloth after keratinase washing. (**d**) Protein-fouling cloth after trypsin washing. (**e**) Protein-fouling cloth after complex enzyme detergent. (**f**) Protein-fouling cloth after commercial detergent. Surface image of blood dirty cloth. (**g**) Unwashed blood-dirty cloth. (**h**) Blood-dirty cloth after alkaline protease washing. (**i**) Blood-dirty cloth after keratinase washing. (**j**) Blood-dirty cloth after trypsin washing. (**k**) Blood-dirty cloth after complex enzyme detergent. (**l**) Blood-dirty cloth after commercial detergent. Microscopic images of protein stains after washing protein stains with protease liquid detergent. (**m**) Protein-fouling cloth after washing (500×). (**n**) Blood-dirty cloth after washing (500×).

4. Conclusions

Adding protease to detergents can remove protein stains. The washing performance of different proteases was tested through the protein-fouling cloth model, and the dirty cloth before and after washing was characterized and analyzed. The compound enzyme of the protease selected in the experiment was more efficient in removing protein stains than a single enzyme. Protein stains were disintegrated and peeled off, and stain redeposition resistance was effective. By step-by-step matching with the protease, washing aids such as sodium alginate, sodium gluconate, and sodium carboxymethyl cellulose were selected to further improve the washing effect of the composite protease. Formula A exhibited superior washing performance over common commercial detergents and, therefore, can provide new product solutions for the expansion and development of washing products.

Author Contributions: Conceptualization and methodology: W.Z. and J.X.; validation: W.Z. and J.W.; data analysis: M.Z.; investigation: H.Y.; writing—original draft preparation: W.Z. and J.W.; writing—review & editing: W.Z. and J.X.; funding acquisition: J.X. All authors have read and agreed to the published version of the manuscript.

Funding: This research was funded by the Major Science and Technology Innovation Project in Shandong Province (2019JZZY011001).

Institutional Review Board Statement: Not applicable.

Informed Consent Statement: Not applicable.

Data Availability Statement: All generated and analyzed data used to support the findings of this study are included within the article.

Conflicts of Interest: The authors declare no conflict of interest.

References

1. Mukherjee, A.K.; Adhikari, H.; Rai, S.K. Production of alkaline protease by a thermophilic *Bacillus subtilis* under solid-state fermentation (SSF) condition using Imperata cylindrica grass and potato peel as low-cost medium: Characterization and application of enzyme in detergent formulation. *Biochem. Eng. J.* **2008**, *39*, 353–361. [CrossRef]
2. Vojcic, L.; Pitzler, C.; Korfer, G.; Jakob, F.; Ronny, M.; Maurer, K.H.; Schwaneberg, U. Advances in protease engineering for laundry detergents. *New Biotechnol.* **2015**, *32*, 629–634. [CrossRef] [PubMed]
3. Giri, S.S.; Sukumaran, V.; Sen, S.S.; Oviya, M.; Banu, B.N.; Jena, P.K. Purification and partial characterization of a detergent and oxidizing agent stable alkaline protease from a newly isolated *Bacillus subtilis* VSG-4 of tropical soil. *J. Microbiol.* **2011**, *49*, 455–461. [CrossRef] [PubMed]
4. Al-Ghanayem, A.A.; Joseph, B. Current prospective in using cold-active enzymes as eco-friendly detergent additive. *Appl. Microbiol. Biotechnol.* **2020**, *104*, 2871–2882. [CrossRef] [PubMed]
5. Esposito, T.S.; Marcuschi, M.; Amaral, I.P.; Carvalho, L.B.; Bezerra, R.S. Trypsin from the processing waste of the lane snapper (*Lutjanus synagris*) and its compatibility with oxidants, surfactants and commercial detergents. *J. Agric. Food Chem.* **2010**, *58*, 6433–6439. [CrossRef] [PubMed]
6. Singh, S.; Mangla, J.; Singh, S. Evaluation of *Aspergillus fumigatus* NTCC1222 as a source of enzymes for detergent industry. *Resour. Environ. Sustain.* **2021**, *5*, 100030. [CrossRef]
7. Rai, S.K.; Konwarh, R.; Mukherjee, A.K. Purification, characterization and biotechnological application of an alkaline β-keratinase produced by *Bacillus subtilis* RM-01 in solid-state fermentation using chicken-feather as substrate. *Biochem. Eng. J.* **2009**, *45*, 218–225. [CrossRef]
8. Rai, S.K.; Mukherjee, A.K. Optimization of production of an oxidant and detergent-stable alkaline β-keratinase from *Brevibacillus* sp. strain AS-S10-II: Application of enzyme in laundry detergent formulations and in leather industry. *Biochem. Eng. J.* **2011**, *54*, 47–56. [CrossRef]
9. Rai, S.K.; Roy, J.K.; Mukherjee, A.K. Characterisation of a detergent-stable alkaline protease from a novel thermophilic strain *Paenibacillus tezpurensis* sp. nov. AS-S24-II. *Appl. Microbiol. Biotechnol.* **2010**, *85*, 1437–1450. [CrossRef]
10. Manni, L.; Jellouli, K.; Ghorbel-Bellaaj, O.; Agrebi, R.; Haddar, A.; Sellami-Kamoun, A.; Nasri, M. An oxidant- and solvent-stable protease produced by *Bacillus cereus* SV1: Application in the deproteinization of shrimp wastes and as a laundry detergent additive. *Appl. Biochem. Biotechnol.* **2010**, *160*, 2308–2321. [CrossRef]
11. Ribeiro Cardoso dos Santos, D.M.; Victor dos Santos, C.W.; Barros de Souza, C.; Sarmento de Albuquerque, F.; Marcos dos Santos Oliveira, J.; Vieira Pereira, H.J. Trypsin purified from *Coryphaena hippurus* (common dolphinfish): Purification, characterization, and application in commercial detergents. *Biocatal. Agric. Biotechnol.* **2020**, *25*, 101584. [CrossRef]

12. Paul, T.; Jana, A.; Das, A.; Mandal, A.; Halder, S.K.; Das Mohapatra, P.K.; pati, B.R.; Chandra Mondal, K. Smart cleaning-in-place process through crude keratinase: An eco-friendly cleaning techniques towards dairy industries. *J. Clean. Prod.* **2014**, *76*, 140–153. [CrossRef]

13. Paul, T.; Das, A.; Mandal, A.; Halder, S.K.; Jana, A.; Maity, C.; DasMohapatra, P.K.; Pati, B.R.; Mondal, K.C. An efficient cloth cleaning properties of a crude keratinase combined with detergent: Towards industrial viewpoint. *J. Clean. Prod.* **2014**, *66*, 672–684. [CrossRef]

14. Niyonzima, F.N.; More, S. Purification and properties of detergent-compatible extracellular alkaline protease from *Scopulariopsis* spp. *Prep. Biochem. Biotechnol.* **2014**, *44*, 738–759. [CrossRef]

15. Farias, C.B.B.; Almeida, F.C.G.; Silva, I.A.; Souza, T.C.; Meira, H.M.; Soares da Silva, R.d.C.F.; Luna, J.M.; Santos, V.A.; Converti, A.; Banat, I.M.; et al. Production of green surfactants: Market prospects. *Electron. J. Biotechnol.* **2021**, *51*, 28–39. [CrossRef]

16. Lee, S.; Lee, J.; Yu, H.; Lim, J. Synthesis of environment friendly nonionic surfactants from sugar base and characterization of interfacial properties for detergent application. *J. Ind. Eng. Chem.* **2016**, *38*, 157–166. [CrossRef]

17. Li, Y.; Ma, S.; Fang, X.; Wu, C.; Chen, H.; Zhang, W.; Cao, M.; Liu, J. Water hardness effect on the association and adsorption of cationic cellulose derivative/anionic surfactant mixtures for fabric softener application. *Colloids Surf. A Physicochem. Eng. Asp.* **2021**, *626*, 127031. [CrossRef]

18. Sutanto, S.; van Roosmalen, M.J.E.; Witkamp, G.J. Redeposition in CO_2 textile dry cleaning. *J. Supercrit. Fluids* **2013**, *81*, 183–192. [CrossRef]

19. Oikonomou, E.K.; Christov, N.; Cristobal, G.; Bourgaux, C.; Heux, L.; Boucenna, I.; Berret, J.F. Design of eco-friendly fabric softeners: Structure, rheology and interaction with cellulose nanocrystals. *J. Colloid Interface Sci.* **2018**, *525*, 206–215. [CrossRef]

20. Zambrano, M.C.; Pawlak, J.J.; Daystar, J.; Ankeny, M.; Venditti, R.A. Impact of dyes and finishes on the microfibers released on the laundering of cotton knitted fabrics. *Environ. Pollut.* **2021**, *272*, 115998. [CrossRef]

21. Zhang, J.; Zhang, Y.; Li, W.; Li, X.; Lian, X. Optimizing Detergent Formulation with Enzymes. *J. Surfactants Deterg.* **2014**, *17*, 1059–1067. [CrossRef]

22. Grbavcic, S.; Bezbradica, D.; Izrael-Zivkovic, L.; Avramovic, N.; Milosavic, N.; Karadzic, I.; Knezevic-Jugovic, Z. Production of lipase and protease from an indigenous *Pseudomonas aeruginosa* strain and their evaluation as detergent additives: Compatibility study with detergent ingredients and washing performance. *Bioresour. Technol.* **2011**, *102*, 11226–11233. [CrossRef] [PubMed]

23. Rajkumar, R.; Jayappriyan, K.R.; Rengasamy, R. Purification and characterization of a protease produced by *Bacillus megaterium* RRM2: Application in detergent and dehairing industries. *J. Basic Microbiol.* **2011**, *51*, 614–624. [CrossRef]

24. Khan, M.F.; Kundu, D.; Hazra, C.; Patra, S. A strategic approach of enzyme engineering by attribute ranking and enzyme immobilization on zinc oxide nanoparticles to attain thermostability in mesophilic *Bacillus subtilis* lipase for detergent formulation. *Int. J. Biol. Macromol.* **2019**, *136*, 66–82. [CrossRef] [PubMed]

25. Kalak, T.; Gąsior, K.; Wieczorek, D.; Cierpiszewski, R. Improvement of washing properties of liquid laundry detergents by modification with N-hexadecyl-N,N-dimethyl-3-ammonio-1-propanesulfonate sulfobetaine. *Text. Res. J.* **2020**, *91*, 115–129. [CrossRef]

26. McCutcheon, J.N.; Trimboli, A.R.; Pearl, M.R.; Brooke, H.; Myrick, M.L.; Morgan, S.L. Diffuse reflectance infrared Fourier transform spectroscopy (DRIFTS) detection limits for blood on fabric: Orientation and coating uniformity effects. *Sci. Justice* **2021**, *61*, 603–616. [CrossRef]

27. Gruian, C.; Vanea, E.; Simon, S.; Simon, V. FTIR and XPS studies of protein adsorption onto functionalized bioactive glass. *Biochim. Biophys. Acta* **2012**, *1824*, 873–881. [CrossRef]

28. Rekik, H.; Zarai Jaouadi, N.; Gargouri, F.; Bejar, W.; Frikha, F.; Jmal, N.; Bejar, S.; Jaouadi, B. Production, purification and biochemical characterization of a novel detergent-stable serine alkaline protease from *Bacillus* safensis strain RH12. *Int. J. Biol. Macromol.* **2019**, *121*, 1227–1239. [CrossRef]

29. Saleem, M.; Rehman, A.; Yasmin, R.; Munir, B. Biochemical analysis and investigation on the prospective applications of alkaline protease from a *Bacillus cereus* strain. *Mol. Biol. Rep.* **2012**, *39*, 6399–6408. [CrossRef]

30. Bersi, G.; Vallés, D.; Penna, F.; Cantera, A.M.; Barberis, S. Valorization of fruit by-products of *Bromelia antiacantha* Bertol.: Protease obtaining and its potential as additive for laundry detergents. *Biocatal. Agric. Biotechnol.* **2019**, *18*, 101099. [CrossRef]

31. Emran, M.A.; Ismail, S.A.; Hashem, A.M. Production of detergent stable thermophilic alkaline protease by *Bacillus licheniformis* ALW1. *Biocatal. Agric. Biotechnol.* **2020**, *26*, 101631. [CrossRef]

32. Herrera-Márquez, O.; Fernández-Serrano, M.; Pilamala, M.; Jácome, M.B.; Luzón, G. Stability studies of an amylase and a protease for cleaning processes in the food industry. *Food Bioprod. Process.* **2019**, *117*, 64–73. [CrossRef]

33. Koohsaryan, E.; Anbia, M.; Maghsoodlu, M. Application of zeolites as non-phosphate detergent builders: A review. *J. Environ. Chem. Eng.* **2020**, *8*, 104287. [CrossRef]

34. Cardoso, A.M.; Horn, M.B.; Ferret, L.S.; Azevedo, C.M.N.; Pires, M. Integrated synthesis of zeolites 4A and Na–P1 using coal fly ash for application in the formulation of detergents and swine wastewater treatment. *J. Hazard. Mater.* **2015**, *287*, 69–77. [CrossRef] [PubMed]

35. Lai, C.; Yang, C.; Zhao, Y.; Jia, Y.; Chen, L.; Zhou, C.; Yong, Q. Promoting enzymatic saccharification of organosolv-pretreated poplar sawdust by saponin-rich tea seed waste. *Bioprocess. Biosyst. Eng.* **2020**, *43*, 1999–2007. [CrossRef]

36. Jian, H.L.; Liao, X.X.; Zhu, L.W.; Zhang, W.M.; Jiang, J.X. Synergism and foaming properties in binary mixtures of a biosurfactant derived from *Camellia oleifera* Abel and synthetic surfactants. *J. Colloid Interface Sci.* **2011**, *359*, 487–492. [CrossRef]

37. Lopez, C.G.; Richtering, W. Oscillatory rheology of carboxymethyl cellulose gels: Influence of concentration and pH. *Carbohydr. Polym.* **2021**, *267*, 118117. [CrossRef]
38. Sheng, Y.; Xu, X.; Jiang, W.; Song, Y.; Gan, S.; Zou, H. Application of Oxidized Cornstarch as a Nonphosphoric Detergent Builder. *J. Surfactants Deterg.* **2012**, *15*, 393–398. [CrossRef]
39. Niyonzima, F.N.; More, S.S. Coproduction of detergent compatible bacterial enzymes and stain removal evaluation. *J. Basic Microbiol.* **2015**, *55*, 1149–1158. [CrossRef]
40. Cheng, K.C.; Khoo, Z.S.; Lo, N.W.; Tan, W.J.; Chemmangattuvalappil, N.G. Design and performance optimisation of detergent product containing binary mixture of anionic-nonionic surfactants. *Heliyon* **2020**, *6*, e03861. [CrossRef]

applied sciences

Review

Glucose Isomerase: Functions, Structures, and Applications

Ki Hyun Nam [1,2]

[1] Department of Life Science, Pohang University of Science and Technology, Pohang 37673, Korea; structures@postech.ac.kr

[2] POSTECH Biotech Center, Pohang University of Science and Technology, Pohang 37673, Korea

Abstract: Glucose isomerase (GI, also known as xylose isomerase) reversibly isomerizes D-glucose and D-xylose to D-fructose and D-xylulose, respectively. GI plays an important role in sugar metabolism, fulfilling nutritional requirements in bacteria. In addition, GI is an important industrial enzyme for the production of high-fructose corn syrup and bioethanol. This review introduces the functions, structure, and applications of GI, in addition to presenting updated information on the characteristics of newly discovered GIs and structural information regarding the metal-binding active site of GI and its interaction with the inhibitor xylitol. This review provides an overview of recent advancements in the characterization and engineering of GI, as well as its industrial applications, and will help to guide future research in this field.

Keywords: glucose isomerase; xylose isomerase; high-fructose corn syrup; HFCS; bioethanol; structure

Citation: Nam, K.H. Glucose Isomerase: Functions, Structures, and Applications. *Appl. Sci.* **2022**, *12*, 428. https://doi.org/10.3390/app12010428

Academic Editor: Hidehiko Hirakawa

Received: 30 October 2021
Accepted: 3 December 2021
Published: 3 January 2022

Publisher's Note: MDPI stays neutral with regard to jurisdictional claims in published maps and institutional affiliations.

1. Introduction

Glucose isomerase (GI, EC 5.3.1.5; also known as D-xylose ketol isomerase, xylose isomerase (XI), xylose ketoisomerase, and xylose ketol-isomerase) is widely distributed in bacteria, actinomycetes, fungi, and plants [1,2]. This enzyme is an intramolecular oxidoreductase that can interconvert aldoses and ketoses. GI occupies a pivotal position with respect to its physiological role and commercial applications [3]. It is one of the three most commonly produced industrial enzymes, along with amylase and protease [2,3]. In particular, this enzyme is extensively used in the industrial production of high-fructose corn syrup (HFCS) [2]. Moreover, the bioconversion of xylose to ethanol by GI is important for the production of bioethanol from hemicellulose [4]. In this review, I introduce the molecular function, structural characteristics, and industrial applications of GI and describe the recent crystallographic results related to GI structure.

2. Function

2.1. Substrate Specificity

Glucose/xylose isomerase (GI/XI) is known to catalyze the reversible isomerization of D-glucose and D-xylose to D-fructose and D-xylulose, respectively (Figure 1).

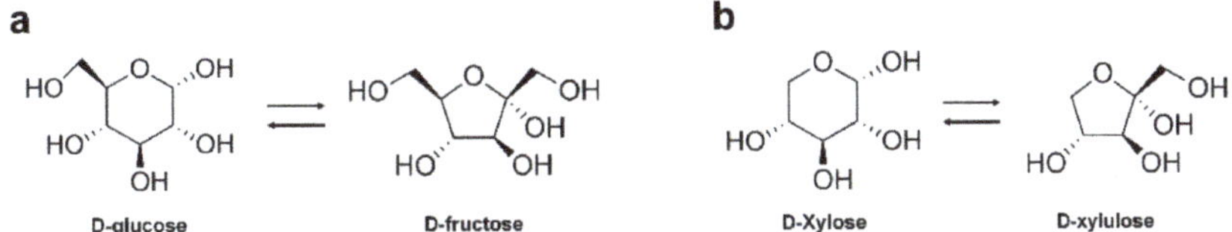

Figure 1. Catalytic reaction of glucose isomerase. Reversible isomerization between (**a**) D-glucose and D-fructose, and (**b**) D-xylose and D-xylulose.

In addition to its natural substrates, GI can isomerize various pentoses, hexoses, sugar alcohols, and sugar phosphates. It exhibits activity against a broad spectrum of sugar substrates, including D-ribose [5], L-rhamnulose [6], L-arabinose [7], and D-allose [8], although the substrate specificity of GIs can vary depending on the source strain or organism [1]. For example, GI from *Streptomyces griseofuscus* S-41 shows isomerase activity on D-glucose (activity 100%) and D-xylose (287%) as well as on D-ribose (20%) [5]. The GI from *Streptomyces rubiginosus* exhibits activity on L-arabinose, in addition to D-glucose and D-xylose [7]. L-arabinose is widely used in food products with a low glycemic index, and in pharmaceutical and chemical industries [9]. The commercially available immobilized GI from *Streptomyces murinus* (SmGI) has previously been used to produce L-rhamnulose (6-deoxy-L-sorbose) from 300 g L^{-1} L-rhamnose in a packed-bed reactor, with a 45% conversion yield [6]. L-rhamnulose plays an important role in sugar metabolism and has wide applications in the flavor industry [6,10]. Furthermore, SmGI has been employed to produce D-allose from D-allulose in a packed bed reactor [11]. At the optimal conditions of pH 8.0 and 60 °C, SmGI produced an average of 150 g/L D-allose over 20 days from 500 g/L D-allulose as the substrate (productivity of 36 g/L/h). D-allose is a rare sugar used as a non-caloric and non-toxic sweetener [8]; it has attracted increasing research interest owing to its beneficial medical properties, including anti-cancer, anti-oxidant, anti-inflammatory, and anti-hypertensive effects [11]. Moreover, SmGI can isomerize D-glucose, D-xylose, D-ribose, and L-rhamnose to D-fructose, D-xylulose, D-ribulose, and L-rhamnulose, respectively [11]. Thus, besides their normal ability to interconvert D-glucose/D-fructose and D-xylose/D-xylulose, GIs can also produce rare sugars that are used in food or medicinal products. GIs from various species could be developed for industrial applications by exploiting their intrinsic substrate specificity for the isomerization for rare sugars.

2.2. Metal Ions

GI requires divalent cations as cofactors for its isomerization activity [12]. Typically, GI is known to exhibit the highest activity in the presence of divalent metal ions such as Mg^{2+}, Mn^{2+} and Co^{2+}; however, the specific metal ion requirement can vary depending on the source of the enzyme. For example, the GI from *Bacillus coagulans* shows the maximum enzyme activity in the presence of Mg^{2+} or Mn^{2+} [13], and Mn^{2+} is the most favorable ion for catalytic activity of the GI from *Piromyces* sp. E2 [14]; in contrast, the GI from the *Streptomyces* strain YT-5 prefers Co^{2+} [15]. Moreover, using a combination of these divalent metal ions has been shown to further enhance the isomerization activity. For example, the activity of EDTA-treated GI from *Streptomyces* sp. CH7 was reportedly increased by 3.6-, 2.8-, and 2.1-fold in the presence of 1 mM Mg^{2+}, Mn^{2+}, and Co^{2+}, respectively, compared to its activity in the absence of any metal ions [16]. However, the enzyme activity increased 6.2-fold and 44.2-fold in the presence of the combinations 1 mM Mg^{2+} and 0.1 mM Co^{2+}, and 10 mM Mg^{2+} and 0.1 mM Co^{2+}, respectively [16]. Thus, during optimization, it is important to measure GI activity not only with one metal ion, but also with a combination of several metal ions. In general, Mg^{2+} and Mn^{2+} act as GI activators, whereas Co^{2+} acts as a protein stabilizer, maintaining the ordered conformation in the quaternary structure of GI [1].

2.3. Reaction Mechanism

The catalytic reaction of GI occurs in three major steps: (i) ring opening, (ii) isomerization, and (iii) ring closure [17] (Figure 2). The action of GI has been attributed to a hydride shift mechanism, based on results from several studies employing chemical modification, X-ray crystallography, and isotope exchange methods [18–20]. Accordingly, it is proposed that the substrate at the active site of GI exists in an open ring state, and a closed ring form product is created after isomerization via a hydride shift from C2 to C1 [18–20]. During this process, a conserved histidine residue catalyzes the proton transfer from O1 to O5. Crystallographic studies have shown that, during its isomerization, xylose initially binds to the enzyme in an open chain conformation. The GI structure contains two sites (M1 and M2) for metal binding. The metal ion at the M1 site binds to O2 and O4 of the xylose molecule, and once bound, the metal ion at the M2 site binds to O1 and O2 in the transition state. These interactions, along with a conserved lysine residue, help catalyze the hydride shift necessary for isomerization [18–20].

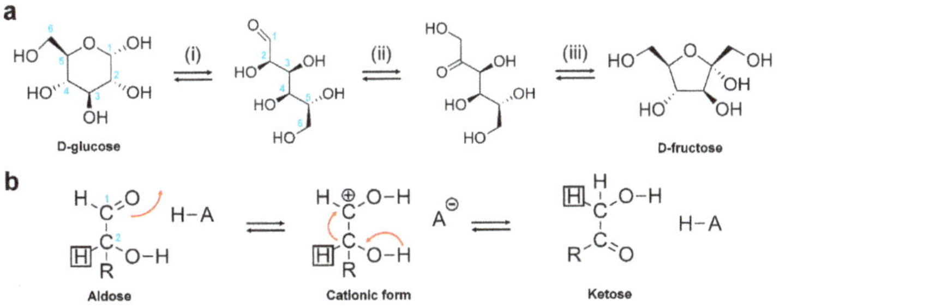

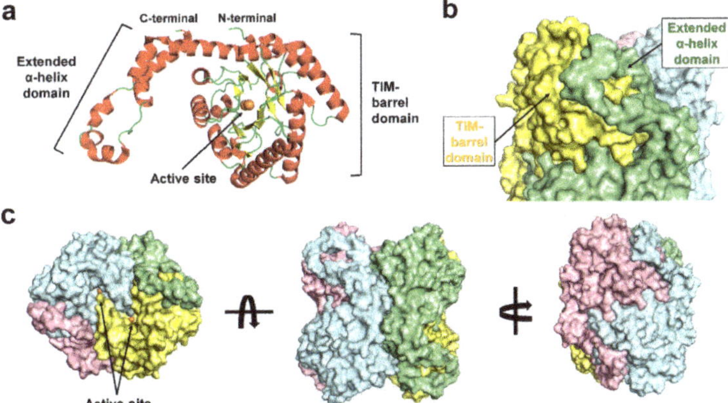

Figure 2. Isomerization mechanism of GI. (**a**) Three major steps involved in the configuration change from D-glucose to D-fructose catalyzed by GI. (**b**) Hydride shift mechanism of GI.

3. Structure

3.1. Overall Structure

The first crystal structure of GI was reported from *Streptomyces rubiginosus* at a 4 Å resolution in 2001 [21]. To date, more than 120 crystal structures of GI/XIs from various species, such as *S. rubiginosus*, *Arthrobacter* sp. NRRL B3728, *Piromyces* sp. E2, *Actinoplanes missouriensis*, *Streptomyces olivochromogenes*, *Streptomyces diastaticus*, *Streptomyces* sp. F-1, *Streptomyces* sp. SK, *Bacteroides thetaiotaomicron* VPI-5482, *Geobacillus stearothermophilus*, *Paenibacillus* sp. R4, *Streptomyces albus*, *S. murinus*, *Thermoanaerobacterium thermosulfurigenes*, *Thermotoga neapolitana*, *Thermus caldophilus*, and *Thermus thermophilus* HB8, have been elucidated and deposited in the Protein Data Bank (PDB, www.rcsb.org). These include more than 100 crystal structures of the GI from *S. rubiginosus*; however, many of these structures have been solved and used only as model samples for the development of various X-ray technologies or studying radiation damage, because the GI from *S. rubiginosus* exhibits high-quality diffraction intensity [22–29].

Structurally, GI consists of a TIM barrel-like domain and an extended α-helix domain (Figure 3A,B) which are assembled into a functional tetramer (Figure 3C) [21]. The substrate-binding pocket is formed by two protomers and the GI active site is located on the TIM-barrel fold (Figure 3C). The GI exists as a tetramer with 222 crystallographic symmetry [30] and includes four active sites for substrate isomerization.

Figure 3. Crystal structure of the GI from *Streptomyces rubiginosus* (SruGI). (**a**) GI protomer consists of a TIM-barrel domain and an extended α-helical domain. (**b**) The extended α-helical domain interacts with the TIM barrel domain from neighboring GI molecule. (**c**) Tetrameric GI shows 222 symmetry.

The typical GI active site contains two sites, M1 and M2, for metal binding. M1 and M2 are referred to as the structural metal and catalytic metal sites, respectively [31], as the metals bound at the M1 and M2 sites are involved in the substrate binding and isomerization mechanism, respectively. In *S. rubiginosus*, the metal at the M1 site is coordinated by Glu181, Glu217, Asp245, and Asp287, and the metal at the M2 site binds the protein via Glu217, His220, Asp255, and Asp257. These metal-bound amino acids are conserved across the GI family [32,33] (Figure 4).

Most of the crystal structures of GI are bound to the metal ions Mg^{2+}, Mn^{2+}, or Co^{2+}, which are involved in the isomerase activity [1]. However, some GI crystal structures are known to contain biologically less related metal ions in their active sites [34,35]. For example, in the crystal structures of GIs from *S. rubiginosus* (PDB code: 4W4Q) [36], *Paenibacillus* sp. *R4* (PDB code: 6INT) [34], and *Piromyces* (6T8E) [35], Ca^{2+} is located in the M1 and M2 sites of the active site, although enzyme activity in the presence of Ca^{2+} has not been verified. In addition, in some crystal structures, the metal ion was placed in a position other than at the M1 and M2 sites [37]. Thus, metal ions in the crystal structures of GI require verification through biochemical experiments or X-ray analysis; additionally, precise analysis of electron density is required to confirm their position and involvement in GI substrate binding and activity.

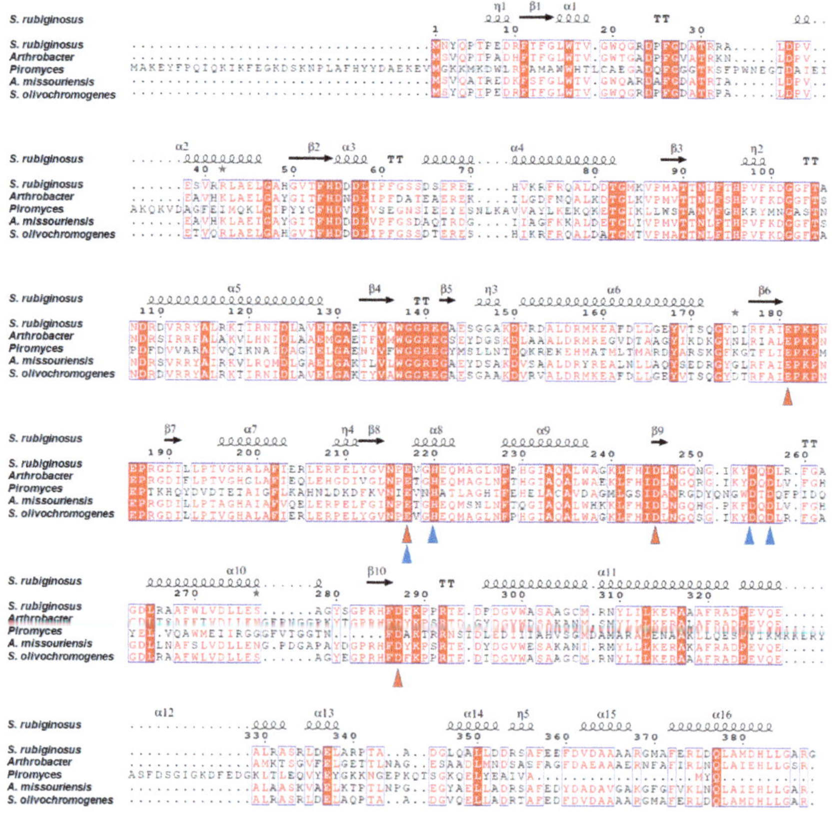

Figure 4. Amino acid sequence alignment of GIs from *Streptomyces rubiginosus* (UniProt Accession: P24300), *Arthrobacter* sp. NRRL B3728 (P12070), *Piromyces* sp. E2 (Q9P8C9), *Actinoplanes missouriensis* (P12851), and *Streptomyces olivochromogenes* (P15587). Residues involved in metal binding at the M1 and M2 sites are indicated by red and blue triangles, respectively. Alignment was performed using Clustal Omega [38] and displayed using Espript [39].

3.2. Metal-Binding State at the GI Active Site

With regard to metal binding, the active site of GI can exist in three different states: (i) two-metal-binding state, (ii) one-metal-binding state, and (iii) metal free-state (Figure 5) [23,33,40]. Although GI crystal structures exhibit different metal-binding states at their active site, the overall GI fold is almost similar, indicating that metal binding does not influence the overall protein folding [40].

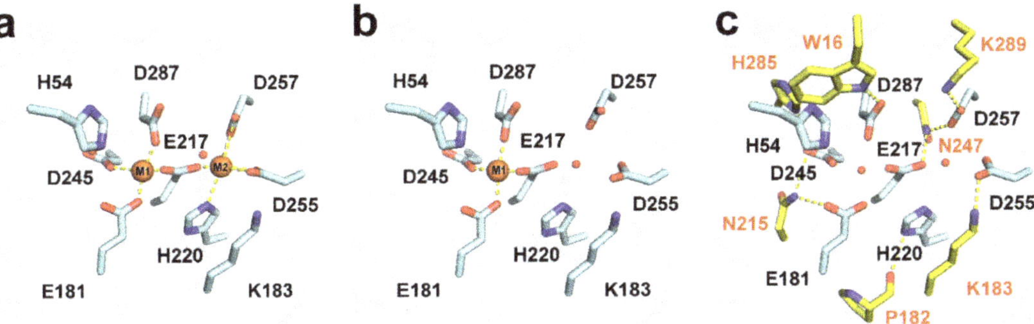

Figure 5. Three different metal-binding states in the active site of the GI from *Streptomyces rubiginosus* (SruGI). (**a**) Two-metal-binding mode (PDB code: 6IRK), (**b**) one-metal-binding mode (PDB code: 5Y4I), and (**c**) metal-free state of SruGI (PDB code: 7CJP).

In the two-metal- binding state of GI, the M1 and M2 sites at active site are occupied by divalent metal ions such as Mg^{2+}, Mn^{2+}, and Co^{2+}, and the differences in enzyme activity occur depending on the metal occupying the active site [41,42]. In the crystal structure of the GI from *S. rubiginosus*, determined via cryo-crystallography, the metal ion at the M1 site is coordinated by Glu181, Glu217, Asp245, and Asp287, as well as the glycerol or ethylene glycol molecule, which is used as a cryoprotectant [33,40]. Meanwhile, in the room-temperature structure of *S. rubiginosus* GI, the metal ion at the M1 site is coordinated by the conserved metal-binding residues and water molecules [23,37], and the metal ion at the M2 site is coordinated by the conserved Glu217, His220, Asp255 (binate interaction), Asp257, and water molecules. The distance between the metal-binding residues and the metal is shorter at the M1 site than at the M2 site, indicating that the M1 site has tighter metal–protein interactions than the M2 site.

In the one-metal-binding mode of GI, the metal ion exists only at the M1 site involved in substrate recognition [33], indicating the higher affinity of the M1 site for metal binding than the M2 site. This result is consistent with the tighter interaction previously observed between the metal-binding residues and the metal at the M1 site, than the interaction at the M2 site [37]. In the one-metal-binding mode, the sugar substrate or inhibitor can interact with the metal ion at the M1 site [33]. However, isomerization activity through a hydride shift mechanism cannot occur due to the absence of a metal ion at the M2 site, leading to an inactive state of GI.

In the metal-free state of GI, no metal is found at both the M1 and M2 sites [40]. In this state, the substrate cannot bind to the active site because neither the metal binding to the M2 site involved in the isomerization activity nor the metal of the M1 site to which the substrate binds exist. In the metal-free state, conformation changes in conserved metal-binding residues occur within 1 Å despite the absence of metal ions at the two metal sites [40]. This is attributed to stabilization of the position of the metal-binding residues by the neighboring residues around the active sites (Glu181 *-Asn215, Glu217 *-Asn247, His220 *-Pro182, Asp245 *-Asn215, Asp245 *-His285, Asp255 *-Lys183, Asp257 *-Asn247, Asp257 *-Lys289, and Asp287 *-Trp16; metal-binding residues are indicated by an asterisk), which helps them maintain their positions without large conformational changes [40]. Accordingly, the metal-free state exhibits a minimum open configuration for the metal to perceive and bind to the sites [40]. Based on the existing biochemical knowledge, this study proposed that, when combining with GI in a metal-free state, the metal first binds to the M1 site and then is filled in the M2 site.

3.3. Xylitol Binding to the Active Site of GI

Structure-based inhibitor studies of GI are helpful for industrial applications such as HGFS and bioethanol production [37], by identifying the parameters of enzyme engineering to prevent product inhibition. The isomerase activity of GI can be inhibited by some divalent cations such as Ca^{2+}, Cu^{2+}, Zn^{2+}, Ni^{2+}, Ag^{2+}, and Hg^{2+} [1]. In addition, molecules such as xylitol, arabitol, sorbitol, mannitol, lyxose, and Tris also inhibit GI activity [43]. However, the mechanism of GI inhibition by these compounds remains largely unclear.

Among these compounds, the inhibition of the activity of GI by xylitol, which is a structural analogue of xylose, has been well studied [44,45]. To date, six crystal structures of xylitol-bound GI have been deposited in PDB (PDB code 1XIG, 2XIS, 3GNX, 4DUO, 5Y4J, and 7DFK) [31,33,37,46,47]. All these structures reveal that the xylitol molecule binds at the M1 site in the active site of GI. Three oxygen (O2, O3, and O4) atoms from xylitol interact with the metal ion at the M1 site, and the metal ion is stabilized by octahedral coordination, by the metal-interacting residues of the enzyme, and the xylitol molecules [31,33,37,46,47]. Therefore, xylitol binding in the M1 site prevents substrate access to the metal ion. The crystal structure of one-metal-binding mode of GI shows that the metal ion at the M2 site is not essential to facilitate xylitol binding to the M1 site [33]. Furthermore, xylitol binding to the M1 site stabilizes the geometry of the metal-binding sites.

Meanwhile, the xylitol-bound GI structure shows a high B-factor value of the metal ion at the M2 site, although the reason for this observation remains unknown. A recently obtained room-temperature structure of xylitol-bound GI via serial crystallography revealed a correlation between xylitol binding and low metal occupancy at the M2 site [37]. In native GI, the M1 site shows a distorted octahedral coordination, but it is geometrically stabilized upon xylitol binding to the metal at the M1 site [37] (Figure 6A). This causes a rearrangement of the conformation of metal-binding residues in the M2 site (Figure 6C). This induces the distortion of metal coordination at the M2 site, leading to the release of the unstable coordinated metal ion from the M2 site [37]. This metal release induces further conformational changes in the neighboring residues (Figure 6C). Since the metal ion at the M2 site involved in isomerization activity is released as a result of xylitol binding, an additional catalytic metal ion has to be added for an isomerization reaction in industrial applications after the release of xylitol.

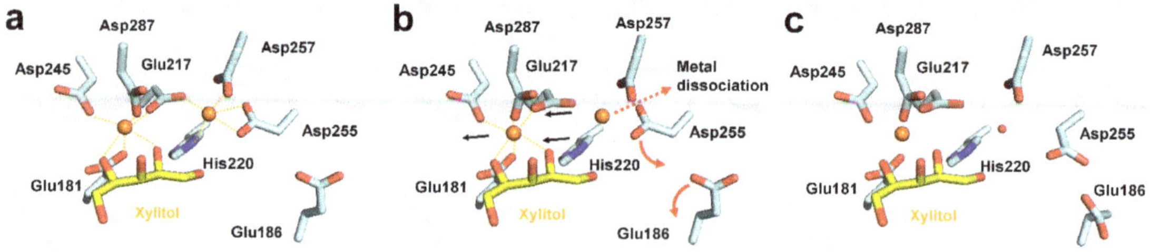

Figure 6. Proposed mechanism of xylitol-induced release of metal at the M2 site of GI. (**a**) Xylitol-bound state of GI. (**b**) Rearrangement of xylitol-binding residues. (**c**) Release of metal ion from M2 site of GI.

4. Application of GI

4.1. High-Fructose Corn Syrup (HFCS)

Owing to its beneficial properties, including a high solubility at low temperature, a lower tendency to crystallize than sucrose, and a high freezing point depression, fructose is widely used in the manufacture of ice cream and frozen desserts to influence taste and texture [48]. Fructose is also used in the production of cake, biscuits, bread, and other confectionery products [49,50]. Moreover, fructose increases the shelf-life of food products, with its high osmotic pressure in solution, making it a better preservative against microbial growth than sucrose syrup [51]. Additionally, fructose is used in the pharmaceutical industry to manufacture diabetes medicine, as it does not influence the blood levels of glucose and insulin [52].

HFCS is an equilibrium mixture of glucose and fructose, which has the advantages of sweetness, low cost, and high solubility [53]. Depending on the fructose content, HFCS can be classified as HFCS-42 (42% fructose, 53% glucose, and 5% polysaccharide), HFCS-55 (55% fructose, 42% glucose, and 3% polysaccharide), and HFCS-90 (90% fructose, 9% glucose and 1% polysaccharide) [54]. Among them, HFCS-55 is the most commonly used, with its commercial production involving several processes such as chromatography, purification, and concentration [55].

HFCS is widely applied in the food, detergent, and pharmaceutical industries [56]. Although glucose can be converted to fructose using a chemical process for HFCS production, this chemical reaction is non-specific and leads to the formation of non-metabolizable sugars [4]. In contrast, GI catalyzes the isomerization of glucose to fructose with excellent specificity, which is critical for the industrial production and application of HFCS [4].

As GI exhibits a reversible isomerization activity, a thermodynamic equilibrium exists in the isomerization reaction between glucose and fructose [57]. The rate of the enzymatic conversion of glucose to fructose can be increased by increasing the reaction temperature; therefore, using a highly thermostable GI capable of sustained operation at high temperatures is critical for one-step HFCS production [57].

Brown et al. reported a GI from the thermophilic eubacterium *Thermotoga maritima* (TmaGI) [58], which is produced when the bacterium is grown in the presence of xylose. TmaGI requires the metal cations Co^{2+} and Mg^{2+} for enzyme activity [58]. This enzyme prefers xylose as a substrate and is most active at pH 6.5–7.5. TmaGI displays the maximum activity at an optimum temperature of 105 to 110 °C, and its half-life is approximately 10 min at 120 °C and pH 7.0. TmaGI is a promising candidate for improving the efficiency of the industrial glucose isomerization process, as it exhibits optimal pH from neutral to slightly acidic, as well as a high thermostability [58].

Deng et al. reported the characterization of GI from *Thermobifida fusca* WSH03-11 (TfuGI) [59]. TfuGI displayed the maximum activity at an optimum temperature of 80 °C, with a half-life of approximately 2 h at 80 °C or 15 h at 70 °C. TfuGI was the most active at pH 10 and retained 95% of its initial activity after incubation at pH 5–10 and 4 °C for 24 h. Analysis of TfuGI enzyme kinetics revealed the K_m and k_{cat} values to be 197 mM and 1688 min^{-1}, respectively [59]. TfuGI was able to convert glucose (45% *w/v*) to fructose, with a maximum conversion yield of 53% at pH 7.5 and 70 °C.

Jia et al. have characterized GIs from *Thermoanaerobacterium xylanolyticum* (TxyGI), *Thermus oshimai* (TosGI), *Geobacillus thermocatenulatus* (GthGI), and *Thermoanaerobacter siderophilus* (TsiGI) [60]. These enzymes were identified using a genome mining approach for their potential application in manufacturing HFCS at elevated temperatures with a low cost of enriching syrups. Among these enzymes, TosGI showed the highest catalytic efficiency toward D-glucose, along with superior thermostability. The optimum temperature of TosGI was 95 °C, and it retained more than 80% of its activity after 48 h at 85 °C in the presence of 20 mM Mn^{2+} [60]. The kinetic parameters K_m and k_{cat}/K_m of TosGI were calculated to be 81.46 mM and 21.77 min^{-1} mM^{-1}, respectively [60]. TosGI achieved a maximum yield of 52.16% for the conversion of D-glucose (400 g/L) to D-fructose at 85 °C [60].

The GI from *Caldicellulosiruptor bescii* (CbeGI), characterized by Dai et al. [57], exhibited the maximum activity at pH 7.0 and 80 °C and retained good thermostability at 85 °C. CbeGI showed affinity for D-glucose, with a K_m of 42.61 mM, and a conversion efficiency up to 57.3% with 3 M D-glucose [57]. The high catalytic efficiency and affinity of CbeGI make it a valuable enzyme for the direct production of 55% HFCS.

Vieille et al. reported the characterization of GI from the hyperthermophile *T. neapolitana* 5068 (TnGI) [61]. The optimal temperature of TnGI was above 95 °C and the optimum pH was 7.1, but the enzyme also showed high activity over a wide pH range. Kinetics studies showed that the catalytic efficiency (k_{cat}/K_m) of TnGI remained constant between 60 and 90 °C; however, the catalytic efficiency decreased between 90 and 98 °C, primarily because of a large increase in K_m [61]. TnGI had a higher turnover number and a lower Km for glucose compared to other thermophilic GIs [61]. Taken together, thermophilic GIs exhibit various activities according to the species of origin and are expected to be extensively applied in industrial HFCS production.

4.2. Ethanol Production

The bioconversion of a renewable biomass into ethanol is attractive in view of the rapid depletion of fossil fuels [62]. GI can catalyze the isomerization of xylose derived from hemicellulosic biomass to xylulose, which can be fermented to ethanol by conventional yeast such as *Saccharomyces cerevisiae*, *Schizosaccharomyces pombe*, and *Candida tropicalis* [63–65]. This property of GI is attractive for the application in bioethanol production because no coenzyme is required and no intermediates are produced during the reaction [66]. However, the typical ethanol production process is plagued by low efficiency, owing to the low conversion yield of xylose to ethanol [67]. The GI enzyme engineering and strain improvement approaches have been explored to improve the conversion yield of xylose to ethanol [1,65,68]. The engineering of yeast strains that exhibit faster xylose metabolism is important in pursuing strain improvement for bioethanol production [69,70].

Ko et al. evaluated the performance of SXA-R2P-E, an engineered isomerase-based xylose-utilizing *S. cerevisiae* strain, for co-fermentation of xylose and glucose to ethanol [65]. SXA-R2P-E produced 50g/L ethanol (yield of 0.43 g ethanol/g sugar) in 72 h via a high-sugar fermentation process (70 g/L glucose and 40 g/L xylose). This strain also produced 18-21 g/L ethanol (yield of 0.43–0.46 g ethanol/g sugar) from acid-pretreated lignocellulosic hydrolysates such as rice straw and hardwood [65].

Seike et al. reported improved xylose fermentation through XI expression in the xylose-utilizing *S. cerevisiae* strain IR-2, which has a deletion of the *GR3* gene (encoding an endogenous xylose reductase) and overexpresses *XK21* (encoding an additional xylulose kinase). Evolutionary engineering was performed in the IR-2 strain to select high-efficiency XIs from eight previously reported XIs derived from various species [68]: *Burkholderia cenocepacia* J2315 [71], *Lachnoclostridium phytofermentans ISDg* (LpXI) [67], *Orpinomyces* sp. ukk1 [72], *Piromyces* sp. E2 [73], *Prevotella ruminicola* TC2-24 [74], *Ruminiclostridium cellulolyticum* H10 [75], *Ruminococcus flavefaciens* 17 [76], and *S. rubiginosus* [77]. Among them, the strain expressing LpXI exhibited the highest D-xylose consumption rate after 72 h of micro-aerobic fermentation on D-glucose and D-xylose mixed medium [68]. Furthermore, to enhance the LpXI catalytic activity, the authors performed random mutagenesis using an error-prone polymerase chain reaction and obtained two enzyme constructs with improved fermentation performances [68]. The SS120 strain, expressing double mutant LpXI (T63I/V162A), produced 53.3 g/L ethanol in 72 h (ethanol yield of approximately 0.44 (g/g-input sugars) from 85 g/L D-glucose and 35 g/L D-xylose [68].

XI enzyme engineering is another approach that has been employed to increase the efficiency of bioethanol production by accelerating xylose metabolism [68,78]. For example, the XI from the fungal strain *Piromyces* sp. E2 (PirXI) is employed for xylose isomerization in engineered *S. cerevisiae* strains but has low activity in vivo. To improve the performance of PirXI, Lee et al. constructed a mutant library by substituting residues around the substrate and metal-binding sites of the enzyme [35]. PirXI variants were obtained by transferring the library to *S. cerevisiae*, followed by in vivo selection for enhanced xylose-supported growth under aerobic and anaerobic conditions. In particular, in the presence of Mg^{2+} or Mn^{2+}, the K_m of PirXI-V270A/A273G variant for xylose was higher, and the k_{cat} was lower than the corresponding values for wild-type PirXI [35].

Thus, further efforts to engineer GI/XI for a high xylose conversion rate through rational engineering or directed evolution are still needed, and applied research should be performed through in vivo studies on improved strains to enable efficient bioethanol production.

5. Perspective

In addition to its biological importance, GI is an industrially important enzyme that is applied in HFCS and bioethanol production. Although biochemical and structural studies have been conducted on GIs from various organisms, many GIs still need to be identified from the protein database. The development of more efficient GI-based industrial processes requires extensive biochemical analysis and engineering of novel GIs. In addition, it will be necessary to analyze not only in vitro enzymatic activity, but also the industrial performance of GI using whole cell systems. On the other hand, structural studies on the inhibition of GI activity by various metals, sugars, and other inhibitors should be conducted to help better modulate GI activity and enhance the enzyme performance during industrial application. Furthermore, it is necessary to understand the dynamical structural changes occurring in GI to aid enzyme engineering efforts, as most GI structures currently collected in cryogenic environments have only provided limited knowledge with regard to flexibility. The recently developed serial crystallography technique provides information regarding the molecular fluidity of proteins at room temperature, and future studies of molecular fluidity based on the room-temperature crystal structures will provide useful information for GI enzyme engineering.

Funding: This work was funded by the National Research Foundation of Korea (NRF) (NRF-2017R1D1A1B03033087, NRF-2017M3A9F6029736 and NRF-2021R1I1A1A01050838) and Korea Initiative for Fostering University of Research and Innovation (KIURI) Program of the NRF (NRF-2020M3H1A1075314).

Institutional Review Board Statement: Not applicable.

Informed Consent Statement: Not applicable.

Data Availability Statement: The data presented in this study are available on request from the corresponding author. The data are not publicly available due to data protection legislation.

Conflicts of Interest: The author declares no conflict of interest.

References

1. Bhosale, S.H.; Rao, M.B.; Deshpande, V.V. Molecular and industrial aspects of glucose isomerase. *Microbiol. Rev.* **1996**, *60*, 280–300. [CrossRef]
2. Kilara, A.; Shahani, K.M. The use of immobilized enzymes in the food industry: A review. *CRC Crit. Rev. Food Sci. Nutr.* **1979**, *12*, 161–198. [CrossRef]
3. Deshpande, V.; Rao, M. Glucose Isomerase. In *Enzyme Technology*; Springer: New York, NY, USA, 2006; pp. 239–252. [CrossRef]
4. Singh, R.S.; Singh, T.; Pandey, A. Microbial Enzymes—An Overview. In *Advances in Enzyme Technology*; Elsevier: Amsterdam, The Netherlands, 2019; pp. 1–40. [CrossRef]
5. Kasumi, T.; Hayashi, K.; Tsumura, N. Purification and Enzymatic Properties of Glucose Isomerase from *Streptomyces griseofuscus*, S-41. *Agric. Biol. Chem.* **2014**, *45*, 619–627. [CrossRef]
6. Kim, Y.-S.; Kim, D.-Y.; Park, C.-S. Production of l-rhamnulose, a rare sugar, from l-rhamnose using commercial immobilized glucose isomerase. *Biocatal. Biotransform.* **2017**, *36*, 417–421. [CrossRef]
7. Langan, P.; Sangha, A.K.; Wymore, T.; Parks, J.M.; Yang, Z.K.; Hanson, B.L.; Fisher, Z.; Mason, S.A.; Blakeley, M.P.; Forsyth, V.T.; et al. L-Arabinose binding, isomerization, and epimerization by D-xylose isomerase: X-ray/neutron crystallographic and molecular simulation study. *Structure* **2014**, *22*, 1287–1300. [CrossRef] [PubMed]
8. Chen, Z.; Chen, J.; Zhang, W.; Zhang, T.; Guang, C.; Mu, W. Recent research on the physiological functions, applications, and biotechnological production of D-allose. *Appl. Microbiol. Biotechnol.* **2018**, *102*, 4269–4278. [CrossRef] [PubMed]
9. Fehér, C. Novel approaches for biotechnological production and application of L-arabinose. *J. Carbohydr. Chem.* **2018**, *37*, 251–284. [CrossRef]

10. Menavuvu, B.T.; Poonperm, W.; Leang, K.; Noguchi, N.; Okada, H.; Morimoto, K.; Granstrom, T.B.; Takada, G.; Izumori, K. Efficient biosynthesis of D-allose from D-psicose by cross-linked recombinant L-rhamnose isomerase: Separation of product by ethanol crystallization. *J. Biosci. Bioeng.* **2006**, *101*, 340–345. [CrossRef] [PubMed]
11. Choi, M.N.; Shin, K.C.; Kim, D.W.; Kim, B.J.; Park, C.S.; Yeom, S.J.; Kim, Y.S. Production of D-Allose From D-Allulose Using Commercial Immobilized Glucose Isomerase. *Front. Bioeng. Biotechnol.* **2021**, *9*, 681253. [CrossRef]
12. Young, J.M.; Schray, K.J.; Mildvan, A.S. Proton magnetic relaxation studies of the interaction of D-xylose and xylitol with D-xylose isomerase. Characterization of metal-enzyme-substrate interactions. *J. Biol. Chem.* **1975**, *250*, 9021–9027. [CrossRef]
13. Yoshimura, S.; Danno, G.-i.; Natake, M. Studies ond-Glucose Isomerizing Activity ofd-Xylose Grown Cells from *Bacillus coagulans*, Strain HN-68. *Agric. Biol. Chem.* **2014**, *30*, 1015–1023. [CrossRef]
14. Lee, M.; Rozeboom, H.J.; de Waal, P.P.; de Jong, R.M.; Dudek, H.M.; Janssen, D.B. Metal Dependence of the Xylose Isomerase from *Piromyces* sp. E2 Explored by Activity Profiling and Protein Crystallography. *Biochemistry* **2017**, *56*, 5991–6005. [CrossRef] [PubMed]
15. Takasaki, Y. Studies on Sugar-isomerizing Enzyme. *Agric. Biol. Chem.* **2014**, *30*, 1247–1253. [CrossRef]
16. Chanitnun, K.; Pinphanichakarn, P. Glucose(xylose) isomerase production by Streptomyces sp. CH7 grown on agricultural residues. *Braz. J. Microbiol.* **2012**, *43*, 1084–1093. [CrossRef] [PubMed]
17. Kovalevsky, A.Y.; Hanson, L.; Fisher, S.Z.; Mustyakimov, M.; Mason, S.A.; Forsyth, V.T.; Blakeley, M.P.; Keen, D.A.; Wagner, T.; Carrell, H.L.; et al. Metal ion roles and the movement of hydrogen during reaction catalyzed by D-xylose isomerase: A joint x-ray and neutron diffraction study. *Structure* **2010**, *18*, 688–699. [CrossRef] [PubMed]
18. Lee, C.Y.; Bagdasarian, M.; Meng, M.H.; Zeikus, J.G. Catalytic mechanism of xylose (glucose) isomerase from Clostridium thermosulfurogenes. Characterization of the structural gene and function of active site histidine. *J. Biol. Chem.* **1990**, *265*, 19082–19090. [CrossRef]
19. Collyer, C.A.; Henrick, K.; Blow, D.M. Mechanism for aldose-ketose interconversion by D-xylose isomerase involving ring opening followed by a 1,2-hydride shift. *J. Mol. Biol.* **1990**, *212*, 211–235. [CrossRef]
20. Collyer, C.A.; Blow, D.M. Observations of reaction intermediates and the mechanism of aldose-ketose interconversion by D-xylose isomerase. *Proc. Natl. Acad. Sci. USA* **1990**, *87*, 1362–1366. [CrossRef]
21. Carrell, H.L.; Rubin, B.H.; Hurley, T.J.; Glusker, J.P. X-ray crystal structure of D-xylose isomerase at 4 Å resolution. *J. Biol. Chem.* **1984**, *259*, 3230–3236. [CrossRef]
22. Taberman, H.; Bury, C.S.; van der Woerd, M.J.; Snell, E.H.; Garman, E.F. Structural knowledge or X-ray damage? A case study on xylose isomerase illustrating both. *J. Synchrotron Radiat.* **2019**, *26*, 931–944. [CrossRef]
23. Lee, D.; Baek, S.; Park, J.; Lee, K.; Kim, J.; Lee, S.J.; Chung, W.K.; Lee, J.L.; Cho, Y.; Nam, K.H. Nylon mesh-based sample holder for fixed-target serial femtosecond crystallography. *Sci. Rep.* **2019**, *9*, 6971. [CrossRef] [PubMed]
24. Lee, K.; Lee, D.; Baek, S.; Park, J.; Lee, S.J.; Park, S.; Chung, W.K.; Lee, J.L.; Cho, H.S.; Cho, Y.; et al. Viscous-medium-based crystal support in a sample holder for fixed-target serial femtosecond crystallography. *J. Appl. Crystallogr.* **2020**, *53*, 1051–1059. [CrossRef]
25. Nam, K.H. Stable sample delivery in viscous media via a capillary for serial crystallography. *J. Appl. Crystallogr.* **2020**, *53*, 45–50. [CrossRef]
26. Nam, K.H. Shortening injection matrix for serial crystallography. *Sci. Rep.* **2020**, *10*, 107. [CrossRef] [PubMed]
27. Nam, K.H. Polysaccharide-Based Injection Matrix for Serial Crystallography. *Int. J. Mol. Sci.* **2020**, *21*, 3332. [CrossRef] [PubMed]
28. Nam, K.H. Lard Injection Matrix for Serial Crystallography. *Int. J. Mol. Sci.* **2020**, *21*, 5977. [CrossRef] [PubMed]
29. Park, S.Y.; Choi, H.; Eo, C.; Cho, Y.; Nam, K.H. Fixed-Target Serial Synchrotron Crystallography Using Nylon Mesh and Enclosed Film-Based Sample Holder. *Crystals* **2020**, *10*, 803. [CrossRef]
30. Dauter, Z.; Dauter, M.; Hemker, J.; Witzel, H.; Wilson, K.S. Crystallisation and preliminary analysis of glucose isomerase from *Streptomyces albus. FEBS Lett.* **1989**, *247*, 1–8. [CrossRef]
31. Whitlow, M.; Howard, A.J.; Finzel, B.C.; Poulos, T.L.; Winborne, E.; Gilliland, G.L. A metal-mediated hydride shift mechanism for xylose isomerase based on the 1.6 Å *Streptomyces rubiginosus* structures with xylitol and D-xylose. *Proteins* **1991**, *9*, 153–173. [CrossRef]
32. Carrell, H.L.; Glusker, J.P.; Burger, V.; Manfre, F.; Tritsch, D.; Biellmann, J.F. X-ray analysis of D-xylose isomerase at 1.9 A: Native enzyme in complex with substrate and with a mechanism-designed inactivator. *Proc. Natl. Acad. Sci. USA* **1989**, *86*, 4440–4444. [CrossRef]
33. Bae, J.E.; Kim, I.J.; Nam, K.H. Crystal structure of glucose isomerase in complex with xylitol inhibitor in one metal binding mode. *Biochem. Biophys. Res. Commun.* **2017**, *493*, 666–670. [CrossRef]
34. Park, S.H.; Kwon, S.; Lee, C.W.; Kim, C.M.; Jeong, C.S.; Kim, K.J.; Hong, J.W.; Kim, H.J.; Park, H.H.; Lee, J.H. Crystal Structure and Functional Characterization of a Xylose Isomerase (PbXI) from the Psychrophilic Soil Microorganism, *Paenibacillus* sp. *J. Microbiol. Biotechnol.* **2019**, *29*, 244–255. [CrossRef]
35. Lee, M.; Rozeboom, H.J.; Keuning, E.; de Waal, P.; Janssen, D.B. Structure-based directed evolution improves *S. cerevisiae* growth on xylose by influencing in vivo enzyme performance. *Biotechnol. Biofuels* **2020**, *13*, 5. [CrossRef]
36. Sugahara, M.; Mizohata, E.; Nango, E.; Suzuki, M.; Tanaka, T.; Masuda, T.; Tanaka, R.; Shimamura, T.; Tanaka, Y.; Suno, C.; et al. Grease matrix as a versatile carrier of proteins for serial crystallography. *Nat. Methods* **2015**, *12*, 61–63. [CrossRef]
37. Nam, K.H. Room-Temperature Structure of Xylitol-Bound Glucose Isomerase by Serial Crystallography: Xylitol Binding in the M1 Site Induces Release of Metal Bound in the M2 Site. *Int. J. Mol. Sci.* **2021**, *22*, 3892. [CrossRef]

38. Sievers, F.; Wilm, A.; Dineen, D.; Gibson, T.J.; Karplus, K.; Li, W.; Lopez, R.; McWilliam, H.; Remmert, M.; Soding, J.; et al. Fast, scalable generation of high-quality protein multiple sequence alignments using Clustal Omega. *Mol. Syst. Biol.* **2011**, *7*, 539. [CrossRef]

39. Gouet, P.; Courcelle, E.; Stuart, D.I.; Metoz, F. ESPript: Analysis of multiple sequence alignments in PostScript. *Bioinformatics* **1999**, *15*, 305–308. [CrossRef]

40. Nam, K.H. Crystal structure of the metal-free state of glucose isomerase reveals its minimal open configuration for metal binding. *Biochem. Biophys. Res. Commun.* **2021**, *547*, 69–74. [CrossRef] [PubMed]

41. Sudfeldt, C.; Schaffer, A.; Kagi, J.H.; Bogumil, R.; Schulz, H.P.; Wulff, S.; Witzel, H. Spectroscopic studies on the metal-ion-binding sites of Co^{2+}-substituted D-xylose isomerase from *Streptomyces rubiginosus. Eur. J. Biochem.* **1990**, *193*, 863–871. [CrossRef] [PubMed]

42. Callens, M.; Kerstershilderson, H.; Vangrysperre, W.; Debruyne, C.K. D-Xylose Isomerase from *Streptomyces violaceoruber:* Structural and Catalytic Roles of Bivalent-Metal Ions. *Enzym. Microb. Technol.* **1988**, *10*, 695–700. [CrossRef]

43. Smith, C.A.; Rangarajan, M.; Hartley, B.S. D-Xylose (D-glucose) isomerase from *Arthrobacter* strain N.R.R.L. B3728. Purification and properties. *Biochem. J.* **1991**, *277*, 255–261. [CrossRef] [PubMed]

44. Fenn, T.D.; Ringe, D.; Petsko, G.A. Xylose isomerase in substrate and inhibitor michaelis states: Atomic resolution studies of a metal-mediated hydride shift. *Biochemistry* **2004**, *43*, 6464–6474. [CrossRef] [PubMed]

45. Volkin, D.B.; Klibanov, A.M. Mechanism of thermoinactivation of immobilized glucose isomerase. *Biotechnol. Bioeng.* **1989**, *33*, 1104–1111. [CrossRef]

46. Carrell, H.L.; Hoier, H.; Glusker, J.P. Modes of binding substrates and their analogues to the enzyme D-xylose isomerase. *Acta Crystallogr. D Biol. Crystallogr.* **1994**, *50*, 113–123. [CrossRef]

47. Kovalevsky, A.; Hanson, B.L.; Mason, S.A.; Forsyth, V.T.; Fisher, Z.; Mustyakimov, M.; Blakeley, M.P.; Keen, D.A.; Langan, P. Inhibition of D-xylose isomerase by polyols: Atomic details by joint X-ray/neutron crystallography. *Acta Crystallogr. D Biol. Crystallogr.* **2012**, *68*, 1201–1206. [CrossRef]

48. White, J.S. Straight talk about high-fructose corn syrup: What it is and what it ain't. *Am. J. Clin. Nutr.* **2008**, *88*, 1716S–1721S. [CrossRef]

49. Sahin, A.W.; Zannini, E.; Coffey, A.; Arendt, E.K. Sugar reduction in bakery products: Current strategies and sourdough technology as a potential novel approach. *Food Res. Int.* **2019**, *126*, 108583. [CrossRef]

50. Zargaraan, A.; Kamaliroosta, L.; Yaghoubi, A.S.; Mirmoghtadaie, L. Effect of Substitution of Sugar by High Fructose Corn Syrup on the Physicochemical Properties of Bakery and Dairy Products: A Review. *Nutr. Food Sci.* **2016**, *3*, 3–11. [CrossRef]

51. Yadav, A.K.; Singh, S.V. Osmotic dehydration of fruits and vegetables: A review. *J. Food Sci. Technol.* **2014**, *51*, 1654–1673. [CrossRef] [PubMed]

52. Basciano, H.; Federico, L.; Adeli, K. Fructose, insulin resistance, and metabolic dyslipidemia. *Nutr. Metab.* **2005**, *2*, 5. [CrossRef]

53. Bhasin, S.; Modi, H.A. Optimization of Fermentation Medium for the Production of Glucose Isomerase Using *Streptomyces* sp. SB-P1. *Biotechnol. Res. Int.* **2012**, *2012*, 874152. [CrossRef]

54. Serna-Saldivar, S.O. Maize: Foods from Maize. In *Reference Module in Food Science*; Elsevier: Amsterdam, The Netherlands, 2016. [CrossRef]

55. Liu, Z.Q.; Zheng, W.; Huang, J.F.; Jin, L.Q.; Jia, D.X.; Zhou, H.Y.; Xu, J.M.; Liao, C.J.; Cheng, X.P.; Mao, B.X.; et al. Improvement and characterization of a hyperthermophilic glucose isomerase from *Thermoanaerobacter ethanolicus* and its application in production of high fructose corn syrup. *J. Ind. Microbiol. Biotechnol.* **2015**, *42*, 1091–1103. [CrossRef]

56. Singh, R.S.; Chauhan, K.; Singh, R.P. Enzymatic Approaches for the Synthesis of High Fructose Syrup. In *Plant Biotechnology: Recent Advancements and Developments*; Springer: Singapore, 2017; pp. 189–211. [CrossRef]

57. Dai, C.; Miao, T.; Hai, J.; Xiao, Y.; Li, Y.; Zhao, J.; Qiu, H.; Xu, B. A Novel Glucose Isomerase from *Caldicellulosiruptor bescii* with Great Potentials in the Production of High-Fructose Corn Syrup. *Biomed Res. Int.* **2020**, *2020*, 1871934. [CrossRef]

58. Brown, S.H.; Sjoholm, C.; Kelly, R.M. Purification and characterization of a highly thermostable glucose isomerase produced by the extremely thermophilic eubacterium, *Thermotoga maritima. Biotechnol. Bioeng.* **1993**, *41*, 878–886. [CrossRef]

59. Deng, H.; Chen, S.; Wu, D.; Chen, J.; Wu, J. Heterologous expression and biochemical characterization of glucose isomerase from *Thermobifida fusca. Bioprocess Biosyst. Eng.* **2014**, *37*, 1211–1219. [CrossRef] [PubMed]

60. Jia, D.X.; Zhou, L.; Zheng, Y.G. Properties of a novel thermostable glucose isomerase mined from *Thermus oshimai* and its application to preparation of high fructose corn syrup. *Enzym. Microb. Technol.* **2017**, *99*, 1–8. [CrossRef] [PubMed]

61. Vieille, C.; Hess, J.M.; Kelly, R.M.; Zeikus, J.G. xylA cloning and sequencing and biochemical characterization of xylose isomerase from *Thermotoga neapolitana. Appl. Environ. Microbiol.* **1995**, *61*, 1867–1875. [CrossRef] [PubMed]

62. Canilha, L.; Kumar Chandel, A.; dos Santos Milessi, T.S.; Fernandes Antunes, F.A.; da Costa Freitas, W.L.; das Gracas Almeida Felipe, M.; da Silva, S.S. Bioconversion of sugarcane biomass into ethanol: An overview about composition, pretreatment methods, detoxification of hydrolysates, enzymatic saccharification, and ethanol fermentation. *J. Biomed. Biotechnol.* **2012**, *2012*, 989572. [CrossRef] [PubMed]

63. Chiang, L.C.; Gong, C.S.; Chen, L.F.; Tsao, G.T. d-Xylulose Fermentation to Ethanol by *Saccharomyces cerevisiae. Appl. Environ. Microbiol.* **1981**, *42*, 284–289. [CrossRef]

64. Gong, C.S.; Chen, L.F.; Flickinger, M.C.; Chiang, L.C.; Tsao, G.T. Production of Ethanol from d-Xylose by Using d-Xylose Isomerase and Yeasts. *Appl. Environ. Microbiol.* **1981**, *41*, 430–436. [CrossRef]

65. Ko, J.K.; Um, Y.; Woo, H.M.; Kim, K.H.; Lee, S.M. Ethanol production from lignocellulosic hydrolysates using engineered *Saccharomyces cerevisiae* harboring xylose isomerase-based pathway. *Bioresour. Technol.* **2016**, *209*, 290–296. [CrossRef] [PubMed]
66. Diao, L.; Liu, Y.; Qian, F.; Yang, J.; Jiang, Y.; Yang, S. Construction of fast xylose-fermenting yeast based on industrial ethanol-producing diploid *Saccharomyces cerevisiae* by rational design and adaptive evolution. *BMC Biotechnol.* **2013**, *13*, 110. [CrossRef] [PubMed]
67. Demeke, M.M.; Dietz, H.; Li, Y.; Foulquie-Moreno, M.R.; Mutturi, S.; Deprez, S.; Den Abt, T.; Bonini, B.M.; Liden, G.; Dumortier, F.; et al. Development of a D-xylose fermenting and inhibitor tolerant industrial *Saccharomyces cerevisiae* strain with high performance in lignocellulose hydrolysates using metabolic and evolutionary engineering. *Biotechnol. Biofuels* **2013**, *6*, 89. [CrossRef] [PubMed]
68. Seike, T.; Kobayashi, Y.; Sahara, T.; Ohgiya, S.; Kamagata, Y.; Fujimori, K.E. Molecular evolutionary engineering of xylose isomerase to improve its catalytic activity and performance of micro-aerobic glucose/xylose co-fermentation in *Saccharomyces cerevisiae*. *Biotechnol. Biofuels* **2019**, *12*, 139. [CrossRef]
69. Bracher, J.M.; Martinez-Rodriguez, O.A.; Dekker, W.J.C.; Verhoeven, M.D.; van Maris, A.J.A.; Pronk, J.T. Reassessment of requirements for anaerobic xylose fermentation by engineered, non-evolved *Saccharomyces cerevisiae* strains. *FEMS Yeast Res.* **2019**, *19*, foy104. [CrossRef]
70. Tran Nguyen Hoang, P.; Ko, J.K.; Gong, G.; Um, Y.; Lee, S.M. Genomic and phenotypic characterization of a refactored xylose-utilizing Saccharomyces cerevisiae strain for lignocellulosic biofuel production. *Biotechnol. Biofuels* **2018**, *11*, 268. [CrossRef]
71. Vilela Lde, F.; de Araujo, V.P.; Paredes Rde, S.; Bon, E.P.; Torres, F.A.; Neves, B.C.; Eleutherio, E.C. Enhanced xylose fermentation and ethanol production by engineered *Saccharomyces cerevisiae* strain. *AMB Express* **2015**, *5*, 16. [CrossRef]
72. Tanino, T.; Hotta, A.; Ito, T.; Ishii, J.; Yamada, R.; Hasunuma, T.; Ogino, C.; Ohmura, N.; Ohshima, T.; Kondo, A. Construction of a xylose-metabolizing yeast by genome integration of xylose isomerase gene and investigation of the effect of xylitol on fermentation. *Appl. Microbiol. Biotechnol.* **2010**, *88*, 1215–1221. [CrossRef]
73. Kuyper, M.; Harhangi, H.R.; Stave, A.K.; Winkler, A.A.; Jetten, M.S.; de Laat, W.T.; den Ridder, J.J.; Op den Camp, H.J.; van Dijken, J.P.; Pronk, J.T. High-level functional expression of a fungal xylose isomerase: The key to efficient ethanolic fermentation of xylose by *Saccharomyces cerevisiae*? *FEMS Yeast Res.* **2003**, *4*, 69–78. [CrossRef]
74. Hector, R.E.; Dien, B.S.; Cotta, M.A.; Mertens, J.A. Growth and fermentation of D-xylose by *Saccharomyces cerevisiae* expressing a novel D-xylose isomerase originating from the bacterium *Prevotella ruminicola* TC2-24. *Biotechnol. Biofuels* **2013**, *6*, 84. [CrossRef]
75. Harcus, D.; Dignard, D.; Lepine, G.; Askew, C.; Raymond, M.; Whiteway, M.; Wu, C. Comparative xylose metabolism among the Ascomycetes *C. albicans*, *S. stipitis* and *S. cerevisiae*. *PLoS ONE* **2013**, *8*, e80733. [CrossRef] [PubMed]
76. Aeling, K.A.; Salmon, K.A.; Laplaza, J.M.; Li, L.; Headman, J.R.; Hutagalung, A.H.; Picataggio, S. Co-fermentation of xylose and cellobiose by an engineered *Saccharomyces cerevisiae*. *J. Ind. Microbiol. Biotechnol.* **2012**, *39*, 1597–1604. [CrossRef] [PubMed]
77. Waltman, M.J.; Yang, Z.K.; Langan, P.; Graham, D.E.; Kovalevsky, A. Engineering acidic *Streptomyces rubiginosus* D-xylose isomerase by rational enzyme design. *Protein Eng. Des. Sel.* **2014**, *27*, 59–64. [CrossRef] [PubMed]
78. Lee, S.M.; Jellison, T.; Alper, H.S. Directed evolution of xylose isomerase for improved xylose catabolism and fermentation in the yeast *Saccharomyces cerevisiae*. *Appl. Environ. Microbiol.* **2012**, *78*, 5708–5716. [CrossRef] [PubMed]

Article

L-Fucose Synthesis Using a Halo- and Thermophilic L-Fucose Isomerase from Polyextremophilic *Halothermothrix orenii*

In Jung Kim [1] and Kyoung Heon Kim [1,2,*]

[1] Department of Biotechnology, Graduate School, Korea University, Seoul 02841, Korea; ij0308@korea.ac.kr
[2] Department of Food Bioscience and Technology, College of Life Sciences and Biotechnology, Korea University, Seoul 02841, Korea
*   Correspondence: khekim@korea.ac.kr

Abstract: L-Fucose isomerase (L-FucI)-mediated isomerization is a promising biotechnological approach for synthesizing various rare sugars of industrial significance, including L-fucose. Extremozymes that can retain their functional conformation under extreme conditions, such as high temperature and salinity, offer favorable applications in bioprocesses that accompany harsh conditions. To date, only one thermophilic L-FucI has been characterized for L-fucose synthesis. Here, we report L-FucI from *Halothermothrix orenii* (*Ho*FucI) which exhibits both halophilic and thermophilic properties. When evaluated under various biochemical conditions, *Ho*FucI exhibited optimal activities at 50–60 °C and pH 7 with 0.5–1 M NaCl in the presence of 1 mM Mn^{2+} as a cofactor. The results obtained here show a unique feature of *Ho*FucI as a polyextremozyme, which facilitates the biotechnological production of L-fucose using this enzyme.

Keywords: L-fucose isomerase; L-fucose; L-fuculose; extremophile; halothermophilic bacteria; *Halothermothrix orenii*

Citation: Kim, I.J.; Kim, K.H. L-Fucose Synthesis Using a Halo- and Thermophilic L-Fucose Isomerase from Polyextremophilic *Halothermothrix orenii*. *Appl. Sci.* **2022**, *12*, 4029. https://doi.org/10.3390/app12084029

Academic Editor: Chih-Ching Huang

Received: 1 April 2022
Accepted: 13 April 2022
Published: 15 April 2022

Publisher's Note: MDPI stays neutral with regard to jurisdictional claims in published maps and institutional affiliations.

1. Introduction

L-Fucose isomerase (L-FucI) is a ketol isomerase that mediates the reversible reaction between L-fucose and L-fuculose (Figure 1) [1–3]. Biologically, L-FucI is involved in the initial process of L-fucose catabolism in various bacteria, in which L-fucose obtained from the environment is converted into L-fuculose, eventually leading to the synthesis of L-lactaldehyde, a metabolically important intermediate, through successive reactions [4]. In addition to L-fucose, L-FucI can metabolize D-arabinose due to its structural similarities [1,2,5,6].

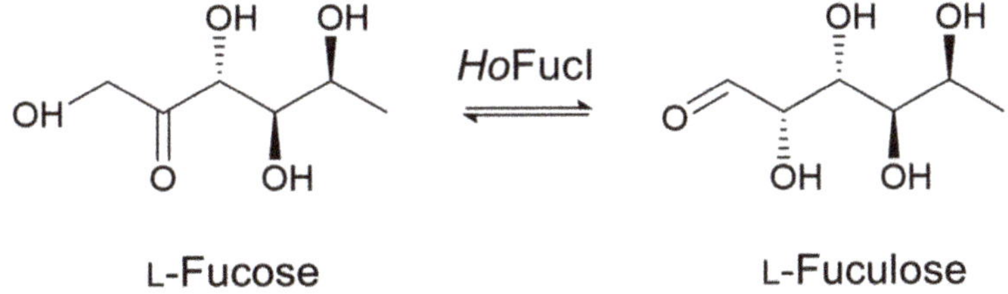

Figure 1. A scheme for interconversion between L-fucose and L-fuculose catalyzed by *Ho*FucI.

In vitro assays have demonstrated that L-FucI-mediated catalysis is not limited to the interconversion between L-fucose–L-fuculose and D-arabinose–D-ribulose; the enzyme is also involved in various isomerization reactions, such as reactions between D-altrose–D-psicose and L-galactose–L-tagatose [6]. Owing to its promiscuous activities, L-FucI is considered to be a biotechnologically useful enzyme for producing rare sugars of industrial significance. Among them, L-fucose is of particular interest due to its diverse bioactive properties, such as anti-inflammatory, antitumor, immune-enhancement, skin-whitening, and skin-moisturization properties, that are beneficial to humans [7–9]. Although widely occurring in nature in the form of bacterial exopolysaccharides, brown seaweed polysaccharides (i.e., fucoidan), and human milk oligosaccharides, it does not meet the industrial demands to directly acquire L-fucose from natural sources [10–12]. Conventional chemical synthesis is also industrially incompatible as it is a time-consuming and labor-intensive multistep process [13,14]. More importantly, the chemical method utilizes toxic reagents, which contradicts the consumer's call for the use of "natural" products. Alternatively, biocatalysts based on either enzymes or whole cells can provide a more efficient and eco-friendly method for L-fucose synthesis. The in vitro L-FucI-based biosynthetic routes for L-fucose known to date are directed towards L-fuculose production as the main precursor, which is subsequently converted into L-fucose by L-FucI [15,16]. Although it is more expensive than L-fucose, strategies to obtain L-fuculose from cheap and common sugars using chemical or enzymatic methods have already been established [15,16]. Thus, L-FucI-mediated L-fucose production utilizing L-fuculose as the substrate is a key target for investigation. Previously, we identified and characterized two L-FucIs from *Raoultella* sp. (*Rd*FucI) [2] and *Thermanaeromonas toyohensis* (*Tt*FucI) [3], demonstrating the applicability of L-FucI for L-fucose synthesis using L-fuculose.

Extremozymes are considered favorable in bioprocesses in the sugar, food, and pharmaceutical fields, which usually accompany high temperatures and high salt concentrations because they are stable and more active under harsh conditions. Such robust enzymes are generally sourced from organisms that have adapted to extreme environments, such as extremely hot and/or saline habitats. To date, three extremo-(thermophilic) L-FucIs (*Tt*FucI, *Cs*FucI from *Caldicellulosiruptor saccharolyticus*, and *Dt*FucI from *Dictyoglomus turgidum*) have been characterized [3,5,6,17]. However, only *Tt*FucI has been evaluated for L-fucose synthesis [3]. Therefore, it is of great importance to identify novel extremo- L-FucIs that could offer great advantages in industrial applications.

Halothermothrix orenii is a true anaerobic polyextremophile isolated from the Tunisian salt lake in the Sahara Desert [18]. Owing to its ability to thrive under dual extreme conditions, that is, high temperature and high salinity, with optimal growth at 60 °C and 1.7 M NaCl, it is classified as a halothermophile [18,19]. Polyextremophilic organisms are very rare and unique in nature, as they have evolved to develop clever strategies to adapt to more than two severe conditions. Accordingly, *H. orenii* could serve as a useful source of enzymes with beneficial features for industrial applications. Indeed, the genome of *H. orenii* revealed the presence of genes encoding enzymes of biotechnological significance, such as those associated with carbohydrate saccharification and metabolism of sugars, including L-fucose [18,19]. In this study, we identified L-FucI from *H. orenii* (*Ho*FucI) and performed its functional characterization for L-fucose synthesis under various biochemical conditions, such as NaCl, temperature, pH, and metal ions, to explore its unique molecular features. The results obtained from this study will not only offer insights into the fundamental knowledge of extremozymes, but also allow the biotechnological production of L-fucose and other valuable rare sugars using *Ho*FucI.

2. Materials and Methods

2.1. Expression and Purification of HoFucI

The gene encoding *Ho*FucI was synthesized (Bioneer, Daejeon, Korea) and cloned into pET28a(+) as an expression vector. The vector containing the synthetic *Ho*FucI gene was transformed into competent *Escherichia coli* BL21 (DE3) cells, which were grown at 37 °C on lysogeny broth (LB)-agar plates containing 50 μg/mL of kanamycin for the selection of transformants. *E. coli* transformants containing the *Ho*FucI gene were cultivated in the LB medium containing 50 μg/mL of kanamycin at 37 °C. When the optical density at 600 nm (OD_{600}) reached 0.6–0.8, the expression of recombinant *Ho*FucI was induced by adding 0.5 mM isopropyl-β-d-1-thiogalactopyranoside to the medium, and the cells were further cultivated at 18 °C and 180 rpm overnight. After centrifugation at $4500 \times g$ at 4 °C for 20 min, the collected cells were resuspended in buffer A (50 mM Tris-HCl, 200 mM NaCl, and 10 mM imidazole; pH 8.0) and disrupted using an ultrasonicator to extract the intracellular proteins. Total cell lysates were centrifuged at $25,188 \times g$ at 4 °C for 30 min, and the supernatant containing the crude protein extract was loaded onto a His-Trap column pre-equilibrated with buffer A (GE Healthcare, Chicago, IL, USA). The *Ho*FucI protein was purified using immobilized metal ion affinity chromatography. After thoroughly washing the column to remove the non-specifically bound proteins with buffer A, the bound protein was eluted with buffer B composed of 50 mM Tris-HCl, 200 mM NaCl, and 300 mM imidazole (pH 8.0). For further purification and determination of the oligomeric state of the native enzyme, the eluted protein fractions were concentrated and subjected to size-exclusion chromatography using a HiPrepTM 16/60 Sephacryl S-200 HR column (GE Healthcare), with a buffer composed of 20 mM sodium phosphate and 200 mM NaCl (pH 7.0) as the mobile phase.

2.2. Enzyme Assay

Generally, enzymatic reactions were performed in a 20 mM sodium phosphate (pH 7.0) buffer containing 1 mM Mn^{2+} as the cofactor and 0.5 M NaCl. Purified *Ho*FucI (7 μg) was incubated with 10 mM L-fuculose at 60 °C for 5 min. The amount of L-fucose obtained from the enzymatic reactions was measured using the K-FUCOSE assay kit (Megazyme, Bray, Co. Wicklow, Ireland), according to the manufacturer's instructions. The experimental data represent the means ± standard deviations. Enzyme activity was represented by specific activity (U/mg) where one unit (U) was defined as the amount of purified *Ho*FucI enzyme required to yield 1 μmol of L-fucose per min from 10 mM L-fucoulose under specified conditions.

2.3. Effects of Various Biochemical Conditions on HoFucI Activity

To investigate the effect of NaCl on the enzymatic activity of *Ho*FucI, 7 μg of *Ho*FucI and 10 mM L-fuculose were incubated at 60 °C in 20 mM sodium phosphate (pH 7.0) buffer in the presence of 1 mM Mn^{2+} for 5 min at different concentrations of NaCl (0, 0.5, 1, 2.5, and 5 M). Temperature and pH profiles were determined under the standard conditions described above. The optimum temperature was examined at various temperatures ranging from 30 to 100 °C with an interval of 10 °C, and the optimum pH was investigated with three buffer systems using 100 mM sodium acetate (pH 4–6), 100 mM sodium phosphate (pH 7–8), and 100 mM glycine-NaOH (pH 9–11). The effect of metal ions on the enzymatic activity of *Ho*FucI was investigated using different concentrations of ethylenediaminetetraacetic acid (EDTA), Mn^{2+}, Mg^{2+}, or Co^{2+} (0, 1, and 10 mM) under standard conditions.

2.4. Phylogenetic Tree

A homolog search was conducted based on the HMMER algorithm with an E-value cutoff of 0.0001 using *Rd*FucI as the query on the ConSurf server (https://consurf.tau.ac.il/ (accessed on 30 March 2022)). Extremely long and obvious fault sequences were discarded. After excluding redundant sequences at the N-terminus, the sequences for L-FucI homologs that were searched and previously characterized were aligned using Clustal Omega [20]. The aligned sequences were then applied to phylogenetic tree building using the ConSurf server. The tree was visualized using FigTree (https://tree.bio.ed.ac.uk/software/figree/ (accessed on 30 March 2022)).

2.5. Sequence Alignment and Homology Modeling

Multiple sequence alignment of L-FucIs was performed based on Clustal Omega [20] and graphically presented by ESPript [21]. The homology model of the *Ho*FucI structure was constructed using SWISS-MODEL (http://swissmodel.expasy.org/ (accessed on 30 March 2022)) with *Rd*FucI (PDB code: 6K1F) as the template, which showed the 65% sequence identity with *HoFucI*. The root mean square deviation (rmsd) between *HoFucI* and *RdFucI* structures was 0.2427 Å. The structures were visualized using the PyMOL software [22].

3. Results

3.1. Oligomeric State of HoFucI

Commonly, native L-FucIs exist as oligomers [2,3,23]. According to crystal structure analysis, L-FucIs form homohexameric structures to be functional for substrate interactions [2,23]. However, the oligomeric states of L-FucIs in a solution can vary, existing as trimers or tetramers, depending on the biochemical conditions [3,6,17]. In this study, size-exclusion chromatography, which is commonly used to examine the enzyme's molecular mass at the native state (i.e. under the non-denaturing condition), was performed to determine the oligomeric state of *HoFucI* (Figure 2). The chromatogram showed a single peak corresponding to *HoFucI* (Figure 2a). By comparing its elution volume with that of standard proteins, the molecular mass of the enzyme was predicted to be approximately 204 kDa, corresponding to a trimer (Figure 2b).

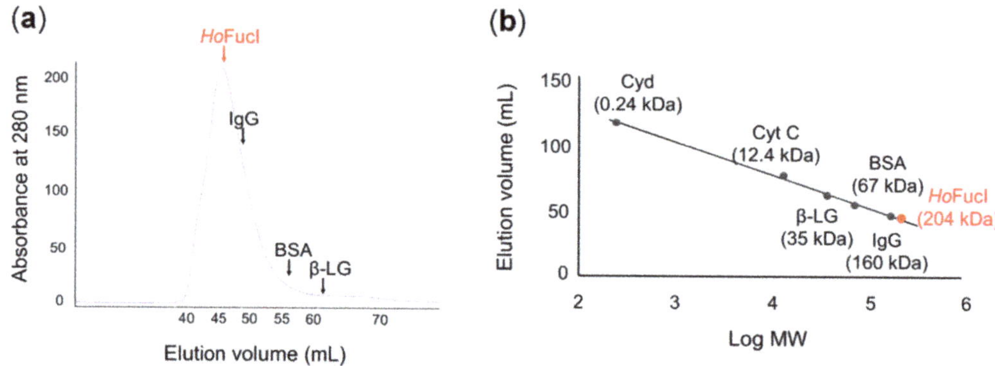

Figure 2. (**a**) Size-exclusion chromatography profile of L-fucose isomerase (L-FucI) from *Halothermothrix orenii* (*HoFucI*). (**b**) Calibration curve obtained using various protein standards (cytidine [Cyd], cytochrome C [Cyt C], β-lactoglobulin [β-LG], bovine serum albumin [BSA], and immunoglobulin G [IgG]) to determine the native molecular mass of *HoFucI*.

3.2. Effect of Salinity on HoFucI Activity

Since the origin of *HoFucI*, *H. orenii*, is a halophilic bacterium that grows optimally at 1.7 M NaCl [18,19]. The effect of salinity on the enzymatic activity of *HoFucI* was evaluated using various concentrations of NaCl from 0 to 5 M (Figure 3). Without NaCl supplementation (control), *HoFucI* showed substantial activity, with a specific activity of 106.2 U/mg. When the NaCl concentration was increased up to 1 M, the isomerization activity of *HoFucI* toward L-fuculose also increased. The maximal activity was exhibited at 0.5 and 1 M with no significant difference ($p > 0.05$), which was approximately 60% higher than that obtained without the addition of NaCl. At concentrations higher than 1 M, the enzyme activity gradually decreased, but was still retained at a higher (at 2.5 M) or similar (at 5.0 M) level compared to the control without the addition of NaCl. Based on these results, *HoFucI* can be classified as a halophilic enzyme that remains active in molar concentrations of the salt.

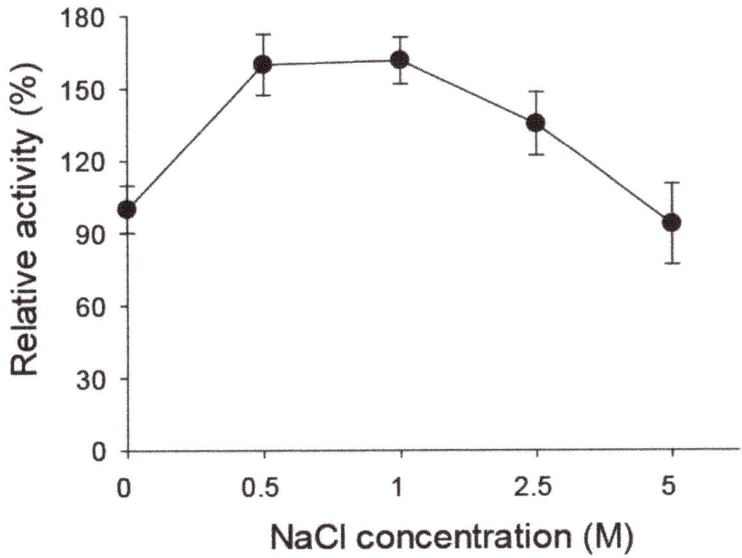

Figure 3. Effect of NaCl concentration on the L-fuculose-isomerizing activity of *Ho*FucI. The enzyme reactions were performed with 7 μg of *Ho*FucI and 10 mM L-fuculose at 60 °C for 5 min in 20 mM sodium phosphate (pH 7.0) buffer in the presence of 1 mM Mn^{2+}. The reaction solution contained various NaCl concentrations ranging from 0 to 5 M. The experimental data represent the means ± standard deviations of three replicates. Relative activity (%) represents the enzyme activity relative to that obtained without supplementation of NaCl, set as 100%.

3.3. Effect of Temperature and pH on HoFucI Activity

When the temperature effect on the activity of *Ho*FucI on L-fuculose was investigated in the presence of 0.5 M NaCl (Figure 4a), the maximum activity was observed at 50 °C and 60 °C with no significant difference ($p > 0.05$), and the enzyme activity sharply decreased at temperatures higher than 60 °C, showing around 60% of relative activity (RA) at 70 °C, but only around 10% activity at 80–100 °C. At 30 and 40 °C, 65% of RA was maintained. This result indicates that *Ho*FucI is a thermophilic enzyme that correlates well with the optimal growth conditions of *H. orenii* at 60 °C, with a range of 42–70 °C [18,19].

When the influence of pH on the enzyme activity of L-fuculose was examined (Figure 4b), the optimal pH was found to be pH 7, and more than 60% of RA was retained at a pH of 8 and 9. At pH values lower than 7 and higher than 9, the activity decreased sharply, showing only approximately 20% of RA at a pH of 6 and 10, and less than 10% of RA at a pH of 4, 5, and 11.

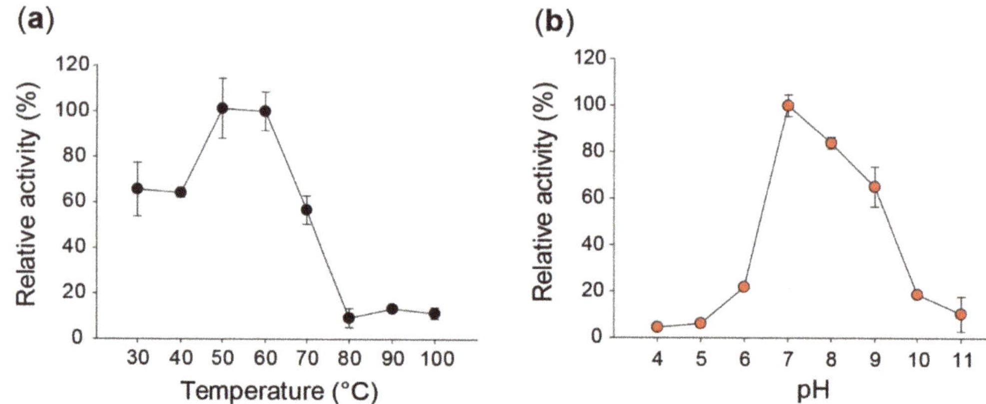

Figure 4. (**a**) Effect of temperature on the L-fuculose-isomerizing activity of *Ho*FucI. Enzyme reactions were performed at various temperatures ranging from 30 to 100 °C with 7 µg of *Ho*FucI and 10 mM L-fuculose for 5 min in 20 mM sodium phosphate (pH 7.0) buffer in the presence of 1 mM Mn^{2+} and 0.5 M NaCl. The experimental data represent the means ± standard deviations. Relative activity (%) represents the enzyme activity relative to that obtained at 60 °C, set as 100%. (**b**) Effect of pH on the L-fuculose-isomerizing activity of *Ho*FucI. Enzyme reactions were performed at various pH ranging from 4 to 11 with 7 µg of *Ho*FucI and 10 mM L-fuculose at 60 °C for 5 min in the presence of 1 mM Mn^{2+} and 0.5 M NaCl. The buffers used were 100 mM sodium acetate (pH 4, 5, and 6), 100 mM sodium phosphate (pH 7 and 8), and 100 mM glycine-NaOH (pH 9, 10, and 11). The experimental data represent the means ± standard deviations. Relative activity (%) represents the enzyme activity relative to that obtained at pH 7, set as 100%.

3.4. Effect of Metal Ions on HoFucI Activity

L-FucIs have a conserved metal-binding site (Figure S1 in Supplementary Materials) and commonly require divalent cations, such as Mn^{2+}, Mg^{2+}, and Co^{2+}, for their catalytic activities [2,5,6]. In particular, Mn^{2+} plays a crucial role in the thermal stability and catalytic activity of L-FucI [3]. In this study, we evaluated the effects of two different concentrations (1 and 10 mM) of Mn^{2+}, Mg^{2+}, and Co^{2+} on the activity of *Ho*FucI. Without any metal supplementation (control), very low activity was observed, while the presence of metal ions boosted the enzyme activity (Table 1). Specifically, upon the addition of 1 mM Mn^{2+} and Co^{2+}, the enzyme activities increased ~5 fold compared to that of the control, but the enzyme activities were lower at 10 mM Mn^{2+} and Co^{2+}. In the case of Mg^{2+}, higher activity was observed at 10 mM Mg^{2+} than at 1 mM, with an RA of 488%. However, EDTA, which is used as a metal chelator, had little effect on enzyme activity.

Table 1. Effect of metal ions on the L-fuculose-isomerizing activity of L-fucose isomerase (L-FucI) from *Halothermothrix orenii* (*Ho*FucI) [1].

Metal Ions	Concentration (mM)		
	0	1	10
EDTA		69.7 ± 24.6	80.8 ± 61.9
Mn^{2+}		555.7 ± 78.7	346.5 ± 7.4
Mg^{2+}	100.0 ± 5.1 [1,2]	252.1 ± 32.1	488.1 ± 19.8
Co^{2+}		523.1 ± 74.2	112.8 ± 19.3

[1] Relative activity (%) represents the enzyme activity relative to that obtained without any supplementation of metal ions, set as 100%. [2] Experimental data represent the means ± standard deviations of three replicates.

3.5. Amino Acid Composition of HoFucI

Halophiles are generally classified into two types depending on their adaptation strategies to external high salt conditions: salt-in and salt-out strategies [24], which are discussed in more detail in the Discussion section. Proteins derived from halophiles that employ the salt-in strategy tend to have a biased amino acid composition, particularly showing a high frequency of acidic residues, such as aspartate and glutamate, and small hydrophobic residues, such as alanine [25,26]. This is often accompanied by a low frequency of basic residues, such as lysine. Accordingly, an acidic signature is a good indicator of the enzymatic properties of halophiles. In this study, the amino acid composition of *HoFucI* was analyzed and compared with those of *RdFucI* and *TtFucI* from non-halophiles, as well as typical proteomes from salt-in halophiles (Table 2). As a result, most of the residues known to be crucial for halostability of salt-in proteins (e.g., alanine, aspartic acid, glutamine, and threonine) were not pronouncedly abundant in *HoFucI* when compared to the non-halophile-derived *RdFucI* and *TtFucI*. In addition, the contents of those residues of *HoFucI* were quite different and low compared to the proteins of salt-in-halophiles. In contrast, its lysine content, which is generally present in reduced frequency in salt-in halophiles, was much higher than that of salt-in proteins. Taken together, these results indicate that the amino acid composition of *HoFucI* is distinct from that of typical halophilic proteins.

Table 2. The content of several amino acid residues related to halophilic properties in *HoFucI*, L-FucI from *Raoultella* sp. (*RdFucI*), L-FucI from *Thermanaeromonas toyohensis* (*TtFucI*), and proteomes from general salt-in halophiles and thermophiles [1].

Amino Acid	*Ho*FucI (%)	*Rd*FucI (%)	*Tt*FucI (%)	Halophiles (%)	Thermophiles (%)	Significance
Alanine	8.5 [2]	11.0	9.7	10.0	8.0	High frequency in halophiles
Aspartic acid	5.3	5.4	5.0	6.0	4.5	Important for halostability
Histidine	1.8	2.7	1.7	2.0	1.8	Low frequency in thermophiles
Lysine	5.3	4.2	4.3	3.0	6.5	Low frequency in halophiles
Glutamine	3.0	4.1	2.8	5.0	3.0	Important for halostability, but low frequency in thermophiles
Glutamate	7.1	6.9	7.0	6.8	7.2	Important for halostability
Threonine	5.5	5.4	5.7	6.2	5.0	Important for halostability, but low frequency in thermophiles

[1] Amino acid compositions of proteomes from general halophiles and thermophiles were adopted from studies by Mavromatis et al. and Bhattacharya et al. [18,19]. [2] The content of each amino acid is expressed as a percentage of the total number of amino acids.

3.6. Phylogenetic Tree of L-FucIs

To date, no phylogenetic tree has been constructed for L-FucI. The phylogenetic tree built in our study showed that many separate clusters of L-FucI sequences were widely distributed among bacteria (Figure 5). Among the characterized L-FucIs, L-FucI from *E. coli* (*EcFucI*), L-FucI from *Klebsiella pneumoniae* (*KpFucI*), and *RdFucI* belonging to the *Enterobacteriaceae* family clustered together. The known thermophilic L-FucIs, such as *DtFucI*, *CsFucI*, and *TtFucI*, were also clustered together as disparate groups from *EcFucI*, *KpFucI*, and *RdFucI*, displaying high phylogenetic relatedness, which indicates their potential functional similarities. Notably, *HoFucI* was positioned in the same cluster as thermophilic L-FucIs, implying that its molecular properties might be similar to those of thermophilic L-FucIs.

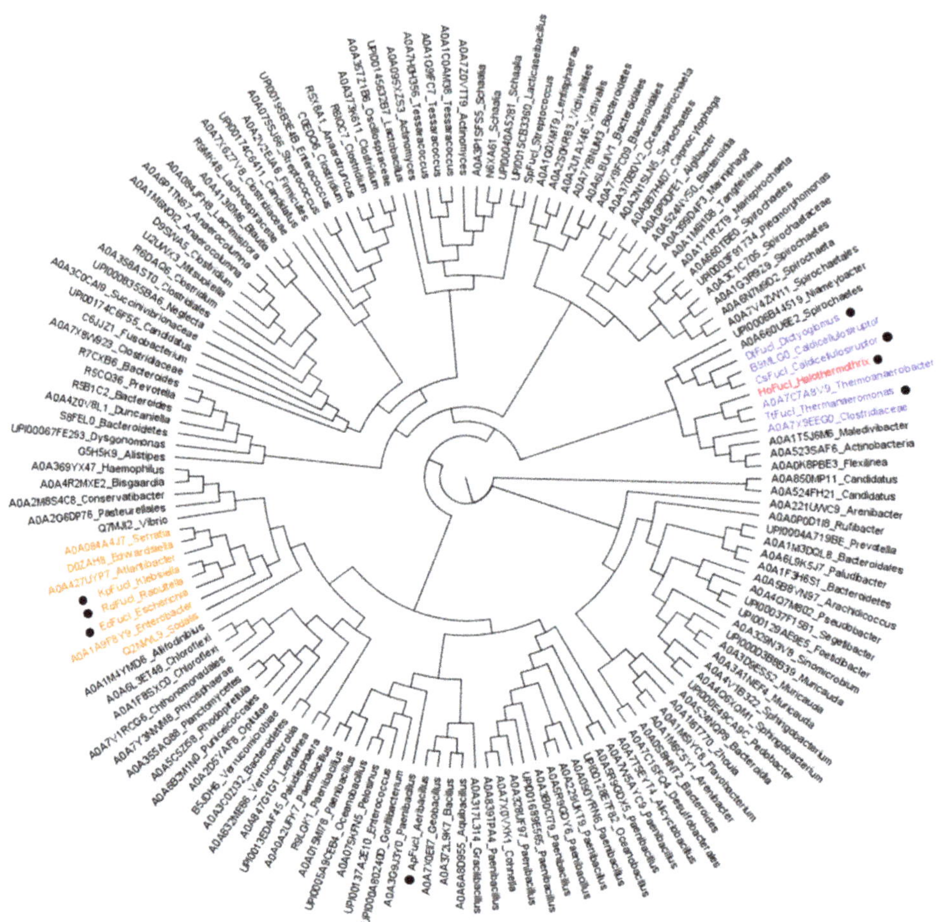

Figure 5. Phylogenetic tree of the L-FucI family. The identities of individual L-FucI protein sequences are indicated by their UniProt accession numbers (https://www.uniprot.org (accessed on 30 March 2022)). The sequences belonging to the *Enterobacteriaceae* family, clustered together with characterized L-FucI from *Escherichia coli* (*Ec*FucI), L-FucI from *Klebsiella pneumoniae* (*Kp*FucI), and L-FucI from *Raoultella* sp. (*Rd*FucI), are highlighted in orange. The sequences from thermophiles, including L-FucI from *Caldicellulosiruptor saccharolyticus* (*Cs*FucI), L-FucI from *Dictyoglomus turgidum* (*Dt*FucI), L-FucI from *Thermanaeromonas toyohensis* (*Tt*FucI), and *Ho*FucI, are highlighted in blue. Functionally characterized L-FucI enzymes were labeled with black circles.

3.7. Comparison of the Specific Activities of L-FucIs against L-Fuculose

The specific activity of *Ho*FucI was compared to those of *Rd*FucI and *Tt*FucI, the previously reported L-FucIs investigated using L-fuculose as the substrate (Table 3). The specific activity of *Ho*FucI obtained without any NaCl (106.2 U/mg) was lower than that of *Tt*FucI (199.8 U/mg) by 1.9 fold, but only 1.3 fold lower in the presence of 0.5 M NaCl (151.1 U/mg). *Ho*FucI activity, both in the absence and presence of the salt, was higher than that of *Rd*FucI (64.0 U/mg). Although its activity was lower than that of *Tt*FucI, the feature of *Ho*FucI that handles the two harsh conditions simultaneously is unique. The key factor leading to the differences in the activities of L-FucIs could be related to the binding affinity and/or geometry of the coordinated metal cofactor complex with the substrate. In addition, the architecture of the substrate recognition pocket and loop confirmation could lead to the difference in L-FucI's activity and substrate specificity [2,3].

Table 3. Comparison of the specific activities of L-FucIs against L-fuculose [1,2].

	Specific Activity (U/mg)	Condition
*Ho*FucI [this study]	106.2 ± 8.7	60 °C, pH 7, w/o NaCl
	151.1 ± 11.9	60 °C, pH 7, w/0.5 M NaCl
*Rd*FucI [2]	64.0 ± 3.4	40 °C, pH 7, w/o NaCl
*Tt*FucI [3]	199.8 ± 0.3	70 °C, pH 7, w/o NaCl

[1] Various concentrations of L-fuculose (10 mM) were incubated with 7 µg *Ho*FucI in 20 mM sodium phosphate (pH 7) buffer at 60 °C for 5 min in the presence of Mn^{2+}. [2] Experimental data represent the means ± standard deviations in at least three replicates.

4. Discussion

L-FucI-encoding genes are distributed throughout a wide range of L-fucose-metabolizing bacteria, many of which are found in the phyla *Firmicutes, Bacteroidetes, Proteobacteria,* and *Actinobacteria* (Figure 5). L-FucI enzymes originate not only from typical mesophilic bacteria, such as *E. coli* [1,23], but also from unique extremophiles [3,5]. Among extremophiles, *H. orenii* represents one of the rarest cases owing to its dual ability to tolerate both high temperature and salinity [18,19]. Organisms that grow optimally at temperatures higher than 50 °C and 1.5 M NaCl are defined as thermophiles and halophiles, respectively. In this regard, *H. orenii* that shows optimal growth at 60 °C and 1.7 M NaCl is classified as a thermohalophile [18,19]. As enzymes derived from such organisms often have distinct features (i.e., high thermostability and halostability) that are beneficial for biotechnological applications [3,5], it is of key interest to investigate their biochemical properties.

Halophiles have evolved to adapt to hypersaline environments by equilibrating intracellular osmotic pressure with their external conditions based on two strategies [24]. One is the "salt-in strategy" that increases the intracellular ion strength up to 3–6 M in response to the external environment; hence, high concentrations of ions like Na^+, K^+, and Cl^-, are accumulated in the cytoplasm [27]. Generally, proteins experience disruption of electrostatic interactions at intracellular molar concentrations of salt, losing their functional conformation [24,28]. In contrast, proteins from organisms utilizing the salt-in strategy are able to maintain their active conformation in the hypersaline cytoplasm. This may be due to the presence of excessive acidic amino acids and negatively charged acidic surfaces, which allow proteins to be sufficiently solvated in limited conditions of water availability by forming electrostatic interactions between the residues and salts [24,26,29]. Consequently, the presence of excessive acidic amino acids (e.g., glutamate and aspartate) over basic amino acids (e.g., lysine and arginine) has been regarded as a unique feature of such halophilic proteins [25,30]. In contrast to the salt-in strategy that increases the intracellular inorganic ions, the "salt-out strategy" exports salts back while accumulating organic solutes in the cytoplasm by synthesizing or acquiring them from the environment to counterbalance the intracellular osmotic pressure [31]. These organic solutes are represented by glycine betaine and ectoine, which are compatible with protein functionality and cellular metabolism [32]. Proteins from halophilic organisms that rely on salt-out strategies do not display acidic signatures.

From the NaCl and temperature profiles in this study, *Ho*FucI was revealed as a poly-extremozyme, evidenced by its thermophilic and halophilic properties in synthesizing L-fucose, with optimal activity at 50–60 °C and 0.5–1 M NaCl (Figures 3 and 4a). This correlates well with the growth behavior of the original organism, *H. orenii*, which has adapted to multi-extreme environments (i.e., hot and salty conditions) [18,19]. It is noteworthy that despite its maximized activity in the presence of a molar salt concentration, *Ho*FucI retained considerable activity even with no or low salt concentrations. Keeping in mind that the halophilic proteins from organisms employing the salt-in strategy tend to be deactivated by low salt concentrations caused by the repulsive electrostatic forces among the excessive acidic amino acids [26], such a phenomenon with *Ho*FucI provides evidence that *H. orenii* does not adopt the salt-in strategy. This is also consistent with the behavior of AmyA (an amylase), one of the few characterized enzymes from the same bacterium, which still exerts 40% activity without salt, while optimal activity is observed at 2 M NaCl [33]. Furthermore, *Ho*FucI is active in a broad range of salinities from 0 to 5 M, which is also consistent with the results of AmyA [33]. Such behaviors of these enzymes could be associated with the extreme seasonal fluctuations of salt in the habitat of *H. orenii* (i.e., Tunisian salt lake of the Sahara desert) [18,19]. Since organisms dependent on the salt-out strategy are generally known to rapidly adjust to fluctuating salinity, the broad activity of *Ho*FucI in saline conditions suggests that *H. orenii* may adopt a salt-out strategy [19].

Our bioinformatics analyses, carried out to better understand the halophilic properties of *Ho*FucI, showed that *Ho*FucI does not strictly follow the general trend of halophilic proteins that have been adapted to high ionic strength (i.e., salt-in strategy). First, there were no pronounced acidic residues, which is an exclusive feature of general halophilic proteins, in *Ho*FucI when the amino acid composition was analyzed (Table 2). Next, because the negative surface electrostatic potentials are characteristic of proteins from organisms that rely on salt-in adaptation [23,34], surface electrostatic potentials were investigated by comparing the model structure of *Ho*FucI with those of its non-halophilic counterparts with known structures, *Ec*FucI and *Rd*FucI (Figure S2 in Supplementary Materials). No distinctive differences in surface electrostatic potentials of the three L-FucI (model) structures were observed, which further proves that the behavior of *Ho*FucI is distinct from those of general halophilic enzymes. AmyA from *H. orenii* also did not show an acidic signature in either the amino acid composition or on its structural surface [33]. Nevertheless, it is too early to conclude that *H. orenii* relies on a salt-out strategy instead of a salt-in strategy because of the lack of related evidence, such as the identification of organic solute-synthetic enzymes [18,19]. Therefore, further studies are required to elucidate the halophilic mechanisms of this bacterium and its enzymes, such as *Ho*FucI. It is noteworthy that proteins from several halophilic organisms belonging to the same order as *H. orenii* (Halanaerobiales) are also very halophilic in enzymatic functioning, but do not follow the general rule regarding the primary and tertiary structures of halozymes, similar to *Ho*FucI in this study [35].

Interestingly, a thermophilic signature was clearly observed in the amino acid composition of *Ho*FucI. Briefly, there was a reduced frequency of thermolabile amino acids, such as histidine, glutamine, and threonine, whose behaviors were quite similar to those of thermophilic *Tt*FucI and proteins from general thermophiles. Further analysis based on the phylogenetic tree also showed that *Ho*FucI is highly evolutionarily related to thermophilic enzymes (Figure 5).

Divalent metal ions are essential cofactors for L-FucIs [2,3,5,6]. These metal ions, in particular Mn^{2+}, coordinated by the three conserved metal-binding residues (Glu337, Asp361, and His528, labeled with *Ec*FucI) in the catalytic center of enzymes, can be platforms for substrate interaction [2,23]. More specifically, the metal cofactors can form a coordination complex with the ligands of the substrate at the O1 and O2 positions, by which L-fucose–L-fuculose interconversion occurs via the enediol intermediate [23]. Sequence alignment showed that *Ho*FucI also contains a conserved metal-binding site (Figure S1 in Supplementary Materials), proving its dependence on metal ions (Table 1). In the cases of Mn^{2+} and Co^{2+}, 1 mM was sufficient to show the maximum activities while a higher concentration of Mg^{2+} (10 mM) was required to obtain a comparable value (Table 1). This could be because although divalent metal ions are expected to be coordinated by metal-binding sites in a similar manner, they might interact with the substrate with different affinities and/or geometries, resulting in the difference in activities. Despite their essential role in the catalytic machinery of L-FucIs, it is necessary to supply additional metal ions to obtain maximal enzyme activity [2,3]. This is well explained by our structural analysis with *Rd*FucI, in which the occupancy of metal ions in the apoenzyme is low, but can be increased by the addition of metal ions [2]. EDTA treatment did not cause any significant difference in the enzyme activity ($p > 0.05$), which further verifies this phenomenon. This was also verified by the fact that the isolated L-FucIs (i.e., apo-form) did not possess any detectable metal ions [5].

5. Conclusions

To the best of our knowledge, the *Ho*FucI studied here represents the first halophilic and thermophilic L-FucI to be identified and characterized. The dual tolerance of *Ho*FucI for heat and salinity, together with its ability to function in a broad range of salt concentrations, is unique and offers attractive biotechnological applications in various industries that often employ harsh conditions, such as food, chemical, and pharmaceutical industries. In addition, since only a few enzymes from *H. orenii* have been reported to date, our study will aid in the further identification and development of useful polyextremozymes from this organism, which is an interesting microbial source for both academic and biotechnological purposes.

Supplementary Materials: The following supporting information can be downloaded at: https://www.mdpi.com/article/10.3390/app12084029/s1, Figure S1: Multiple sequence alignment of l-Fucose isomerase (l-FucI) from Halothermothrix orenii (HoFucI) and previously characterized l-FucIs, including l-FucI from Escherichia coli (EcFucI), l-FucI from Klebsiella pneumoniae (KpFucI), l-FucI from Raoultella sp.(RdFucI), SpFucI, l-FucI from Dictyoglomus turgidum (DtFucI), l-FucI from Caldicellulosiruptor saccharolyticus (CsFucI), l-FucI from Aeribacillus pallidus (ApFucI), and l-FucI from Thermanaeromonas toyohensis (TtFucI); Figure S2: Electrostatic surface potentials of l-FucI from Halothermothrix orenii (HoFucI) (homology model), EcFucI (PDB code: 1FUI), and RdFucI (PDB code: 6K1F).

Author Contributions: Conceptualization, I.J.K. and K.H.K.; Data curation, I.J.K.; Funding acquisition, I.J.K. and K.H.K.; Investigation, I.J.K.; Supervision, K.H.K.; Writing—original draft, I.J.K.; Writing—review and editing, K.H.K. All authors have read and agreed to the published version of the manuscript.

Funding: This work was supported by the Mid-career Researcher Program through the National Research Foundation of Korea (NRF) (2020R1A2B5B02002631) and the Korea Institute of Planning and Evaluation for Technology in Food, Agriculture, Forestry, and Fisheries, funded by the Ministry of Agriculture, Food, and Rural Affairs (32136-05-1-SB010). I.J.K. acknowledges the grant from the NRF (NRF-2020R1A6A3A03039153). This work was also supported by the Korea University Food Safety Hall for the Institute of Biomedical Science and Food Safety and by a Korea University Grant.

Institutional Review Board Statement: Not applicable.

Informed Consent Statement: Not applicable.

Data Availability Statement: Not applicable.

Conflicts of Interest: The authors declare no conflict of interest.

References

1. Green, M.; Cohen, S.S. Enzymatic conversion of L-fucose to L-fuculose. *J. Biol. Chem.* **1956**, *219*, 557–568. [CrossRef]
2. Kim, I.J.; Kim, D.H.; Nam, K.H.; Kim, K.H. Enzymatic synthesis of L-fucose from L-fuculose using a fucose isomerase from *Raoultella* sp. and the biochemical and structural analyses of the enzyme. *Biotechnol. Biofuels* **2019**, *12*, 282. [CrossRef] [PubMed]
3. Kim, I.J.; Kim, K.H. Thermophilic L-fucose isomerase from *Thermanaeromonas toyohensis* for L-fucose synthesis from L-fuculose. *Process Biochem.* **2020**, *96*, 131–137. [CrossRef]
4. Baldomà, L.; Aguilar, J. Metabolism of L-fucose and L-rhamnose in *Escherichia coli*: Aerobic–anaerobic regulation of L-lactaldehyde dissimilation. *J. Bacteriol.* **1988**, *170*, 416–421. [CrossRef] [PubMed]
5. Hong, S.H.; Lim, Y.R.; Kim, Y.S.; Oh, D.K. Molecular characterization of a thermostable L-fucose isomerase from *Dictyoglomus turgidum* that isomerizes L-fucose and D-arabinose. *Biochimie* **2012**, *94*, 1926–1934. [CrossRef] [PubMed]
6. Ju, Y.H.; Oh, D.K. Characterization of a recombinant L-fucose isomerase from *Caldicellulosiruptor saccharolyticus* that isomerizes L-fucose, D-arabinose, D-altrose, and L-galactose. *Biotechnol. Lett.* **2010**, *32*, 299–304. [CrossRef] [PubMed]
7. Péterszegi, G.; Fodil-Bourahla, I.; Robert, A.M.; Robert, L. Pharmacological properties of fucose. Applications in age-related modifications of connective tissues. *Biomed. Pharmacother.* **2003**, *57*, 240–245. [CrossRef]
8. Péterszegi, G.; Isnard, N.; Robert, A.M.; Robert, L. Studies on skin aging. Preparation and properties of fucose-rich oligo- and polysaccharides. Effect on fibroblast proliferation and survival. *Biomed. Pharmacother.* **2003**, *57*, 187–194. [CrossRef]

9. Robert, C.; Robert, A.M.; Robert, L. Effect of a preparation containing a fucose-rich polysaccharide on periorbital wrinkles of human voluntaries. *Skin Res. Technol.* **2005**, *11*, 47–52. [CrossRef]

10. Ale, M.T.; Maruyama, H.; Tamauchi, H.; Mikkelsen, J.D.; Meyer, A.S. Fucose-containing sulfated polysaccharides from brown seaweeds inhibit proliferation of melanoma cells and induce apoptosis by activation of caspase-3 in vitro. *Mar. Drugs* **2011**, *9*, 2605–2621. [CrossRef]

11. Sutherland, I.W. Structural studies on colanic acid, the common exopolysaccharide found in the Enterobacteriaceae, by partial acid hydrolysis. Oligosaccharides from colanic acid. *Biochem. J.* **1969**, *115*, 935–945. [CrossRef] [PubMed]

12. Thurl, S.; Munzert, M.; Henker, J.; Boehm, G.; Müller-Werner, B.; Jelinek, J.; Stahl, B. Variation of human milk oligosaccharides in relation to milk groups and lactational periods. *Br. J. Nutr.* **2010**, *104*, 1261–1271. [CrossRef] [PubMed]

13. Gesson, J.-P.; Jacquesy, J.-C.; Mondon, M.; Petit, P. A short synthesis of L-fucose and analogs from D-mannose. *Tetrahedron Lett.* **1992**, *33*, 3637–3640. [CrossRef]

14. Sarbajna, S.; Das, S.K.; Roy, N. A novel synthesis of L-fucose from D-galactose. *Carbohydr. Res.* **1995**, *270*, 93–96. [CrossRef]

15. Suzuki, S.; Kunihiko, W. Method for Producing L-fuculose and Method for Producing L-fucose. U.S. Patent US/0026504 A1, 1 February 2007. Available online: https://patentimages.storage.googleapis.com/b7/43/40/840ebb8a3ce304/US20070026504A1.pdf (accessed on 30 March 2022).

16. Wong, C.-H.; Alajarin, R.; Moris-Varas, F.; Blanco, O.; Garcia-Junceda, E. Enzymic synthesis of L-fucose and analogs. *J. Org. Chem.* **1995**, *60*, 7360–7363. [CrossRef]

17. Iqbal, M.W.; Riaz, T.; Hassanin, H.A.M.; Ni, D.; Mahmood Khan, I.; Rehman, A.; Mahmood, S.; Adnan, M.; Mu, W. Characterization of a novel D-arabinose isomerase from *Thermanaeromonas toyohensis* and its application for the production of D-ribulose and L-fuculose. *Enzyme Microb. Technol.* **2019**, *131*, 109427. [CrossRef]

18. Mavromatis, K.; Ivanova, N.; Anderson, I.; Lykidis, A.; Hooper, S.D.; Sun, H.; Kunin, V.; Lapidus, A.; Hugenholtz, P.; Patel, B.; et al. Genome analysis of the anaerobic thermohalophilic bacterium *Halothermothrix orenii*. *PLoS ONE* **2009**, *4*, e4192. [CrossRef]

19. Bhattacharya, A.; Pletschke, B.I. Review of the enzymatic machinery of *Halothermothrix orenii* with special reference to industrial applications. *Enzyme Microb. Technol.* **2014**, *55*, 159–169. [CrossRef]

20. Sievers, F.; Higgins, D.G. Clustal omega. *Curr. Protoc. Bioinform.* **2014**, *48*, 3131–3136. [CrossRef]

21. Gouet, P.; Courcelle, E.; Stuart, D.I.; Métoz, F. ESPript: Analysis of multiple sequence alignments in PostScript. *Bioinformatics* **1999**, *15*, 305–308. [CrossRef]

22. DeLano, W.L. Pymol: An open-source molecular graphics tool. *CCP4 Newsl. Prot. Crystallogr.* **2002**, *40*, 82–92.

23. Seemann, J.E.; Schulz, G.E. Structure and mechanism of L-fucose isomerase from *Escherichia coli*. *J. Mol. Biol.* **1997**, *273*, 256–268. [CrossRef] [PubMed]

24. Siglioccolo, A.; Paiardini, A.; Piscitelli, M.; Pascarella, S. Structural adaptation of extreme halophilic proteins through decrease of conserved hydrophobic contact surface. *BMC Struct. Biol.* **2011**, *11*, 50. [CrossRef] [PubMed]

25. Fukuchi, S.; Yoshimune, K.; Wakayama, M.; Moriguchi, M.; Nishikawa, K. Unique amino acid composition of proteins in halophilic bacteria. *J. Mol. Biol.* **2003**, *327*, 347–357. [CrossRef]

26. Ortega, G.; Laín, A.; Tadeo, X.; López-Méndez, B.; Castaño, D.; Millet, O. Halophilic enzyme activation induced by salts. *Sci. Rep.* **2011**, *1*, 6. [CrossRef] [PubMed]

27. Lanyi, J.K. Salt-dependent properties of proteins from extremely halophilic bacteria. *Bacteriol. Rev.* **1974**, *38*, 272–290. [CrossRef] [PubMed]

28. Zaccai, G.; Eisenberg, H. Halophilic proteins and the influence of solvent on protein stabilization. *Trends Biochem. Sci.* **1990**, *15*, 333–337. [CrossRef]

29. Paul, S.; Bag, S.K.; Das, S.; Harvill, E.T.; Dutta, C. Molecular signature of hypersaline adaptation: Insights from genome and proteome composition of halophilic prokaryotes. *Genome Biol.* **2008**, *9*, R70. [CrossRef]

30. Esclapez, J.; Pire, C.; Bautista, V.; Martínez-Espinosa, R.M.; Ferrer, J.; Bonete, M.J. Analysis of acidic surface of *Haloferax mediterranei* glucose dehydrogenase by site-directed mutagenesis. *FEBS Lett.* **2007**, *581*, 837–842. [CrossRef]

31. Roberts, M.F. Organic compatible solutes of halotolerant and halophilic microorganisms. *Saline Syst.* **2005**, *1*, 5. [CrossRef]

32. Weinisch, L.; Kühner, S.; Roth, R.; Grimm, M.; Roth, T.; Netz, D.J.A.; Pierik, A.J.; Filker, S. Identification of osmoadaptive strategies in the halophile, heterotrophic ciliate *Schmidingerothrix salinarum*. *PLoS Biol.* **2018**, *16*, e2003892. [CrossRef] [PubMed]

33. Sivakumar, N.; Li, N.; Tang, J.W.; Patel, B.K.; Swaminathan, K. Crystal structure of AmyA lacks acidic surface and provide insights into protein stability at poly-extreme condition. *FEBS Lett.* **2006**, *580*, 2646–2652. [CrossRef] [PubMed]

34. Pancsa, R.; Kovacs, D.; Tompa, P. Misprediction of structural disorder in Halophiles. *Molecules* **2019**, *24*, 479. [CrossRef] [PubMed]

35. Elevi Bardavid, R.; Oren, A. The amino acid composition of proteins from anaerobic halophilic bacteria of the order Halanaerobiales. *Extremophiles* **2012**, *16*, 567–572. [CrossRef]

Article

Crystal Structure of Human Lysozyme Complexed with *N*-Acetyl-α-D-glucosamine

Ki Hyun Nam [1,2]

[1] Department of Life Science, Pohang University of Science and Technology, Pohang 37673, Korea; structures@postech.ac.kr
[2] POSTECH Biotech Center, Pohang University of Science and Technology, Pohang 37673, Korea

Abstract: Human lysozyme is a natural non-specific immune protein that participates in the immune response of infants against bacterial and viral infections. Lysozyme is a well-known hydrolase that cleaves peptidoglycan in bacterial cell walls. Several crystal structures of human lysozyme have been reported, but little is known regarding how it recognizes sugar molecules. In this study, the crystal structures of human lysozyme in its native and two *N*-acetyl-α-D-glucosamine (α-D-NAG)-bound forms were determined at 1.3 Å and 1.55/1.60 Å resolution, respectively. Human lysozyme formed a typical c-type lysozyme fold and the α-D-NAG molecule was bound to the middle of subsites C and D. The *N*-acetyl and glucosamine groups of α-D-NAG were stabilized by hydrophobic interactions (Val117, Ala126, and Trp127), hydrogen bonds (Asn64, Asn78, Ala126, and Val128), and water bridges. Conformational changes of Arg80, Tyr81, Val128, and Arg131 of human lysozyme were observed due to the interactions of α-D-NAG with the active-site cleft. The binding configuration of α-D-NAG in human lysozyme was distinct compared with that of other sugar-bound lysozymes. Findings from this structural analysis provide a better understanding of the sugar recognition of human lysozyme during the immune response to microbial pathogens.

Keywords: lysozyme; muramidase; *N*-acetylmuramide glycanhydrolase; human; *N*-acetyl-β-D-glucosaminidase; NAG; crystal structure

Citation: Nam, K.H. Crystal Structure of Human Lysozyme Complexed with *N*-Acetyl-α-D-glucosamine. *Appl. Sci.* **2022**, *12*, 4363. https://doi.org/10.3390/app12094363

Academic Editor: Giuseppina Andreotti

Received: 1 April 2022
Accepted: 24 April 2022
Published: 26 April 2022

Publisher's Note: MDPI stays neutral with regard to jurisdictional claims in published maps and institutional affiliations.

1. Introduction

In humans, lysozyme (muramidase or *N*-acetylmuramide glycanhydrolase) is found in several organs, tissues, and secretions. It is also found in the organs and secretions of other vertebrates, invertebrates, plants, and bacteria [1,2]. This enzyme is functionally conserved and exhibits antimicrobial activity that is critical to host defense [3]. Lysozymes are classified into three main families based on their amino acid sequence and biochemical properties. These families include the chicken or conventional type (c-type), goose type (g-type), and invertebrate type (i-type) [2,3]. In addition, phage type, bacterial type, and plant type lysozymes are also found in nature [4–6].

Bacterial cell walls are composed predominantly of peptidoglycan layers. Peptidoglycans consists of alternating units of *N*-acetylglucosamine (NAG or GlcNAc) and *N*-acetylmuramic acid (NAM or MurNAc). The peptide stems are covalently bound to a lactyl group in the NAM moiety [3]. Lysozymes enzymatically cleaves the peptidoglycan in bacterial cell walls by catalyzing the hydrolysis of the β-(1,4) linkages between the NAG and NAM saccharides [7]. These linkages are critical for the resistance of bacteria to osmotic stress, and their hydrolysis results in the loss of cellular membrane integrity and ultimately cell death by lysis [3,8,9]. Due to its broad antibacterial activity against pathogenic bacteria and viruses, lysozyme has been widely used in food, pharmaceutical, and clinical treatments, including various oral health care and antibiotic treatments [10,11].

In humans, lysozyme is highly expressed in breast milk and acts as an important natural non-specific immune protein that participates in the immune response of infants to bacterial [12]. For example, the concentration of lysozyme in the airway surface liquid of humans is estimated to be 20–100 μg/mL, which is sufficient to kill important lung pathogens, such as Gram-negative *Pseudomonas aeruginosa* and Gram-positive *Staphylococcus aureus* [13]. Lysozyme enhances the immune system [14,15], and in addition to antibacterial activity, it may also exhibit antitumor [16] and antiviral [17] activities. However, the molecular mechanism of these activities remains unclear.

Many crystal structures of human lysozyme (hLZM) have been reported, but most crystal structures of hLZM have been used as model samples in X-ray crystallographic studies. For instance, only one crystal structure of saccharide-bound hLZM (PDB ID: 5LSH, tetrasaccharide fragment from the O-chain of lipopolysaccharide from *Klebsiella pneumonia*) [15] has been deposited in the Protein Data Bank (PDB). Accordingly, how hLZM recognizes various types of saccharide molecules has not been fully elucidated. On the other hand, the crystal structure of hLZM complexed with hexa-N-acetyl-chitohexaose has previously been reported [18], but the coordinates and structural factors for this crystal structure are not deposited in PDB. Although the substrate recognition site of hLZM has been studied, it is not known how it recognizes single NAG or NAM sugar molecules constituting the substrate. The observation of single sugar molecule recognition of hLZM is expected to contribute to the initial first sugar molecule recognition when hLZM recognizes the actual substrate. In addition, it is unclear how hLZMs recognize sugars in bacterial cells and to what degree they are evolutionarily conserved with other lysozymes.

To better understand saccharide recognition of hLZM, the crystal structures of native hLZM and hLZM complexed with N-acetyl-α-D-glucosamine (α-D-NAG) were determined at 1.30 Å and 1.55/1.60 Å resolution, respectively. The α-D-NAG molecule was bound to the middle of subsites C and D in the active-site cleft of hLZM. The α-D-NAG molecule was stabilized by conserved hydrophobic interactions, hydrogen bonds, and water bridges. The α-D-NAG binding configuration of hLZM is unique when compared with crystal structure of sugar-bound lysozyme. Results from this study will provide a better understanding of saccharide recognition by hLZM and insight into the human immune response associated with hLZM.

2. Materials and Methods

2.1. Sample Preparation

The hLZM (cat. no. L1667) and N-acetyl-D-glucosamine (cat. no. A4106) were purchased from Sigma-Aldrich (St. Louis, MO, USA). The hLZM was reconstituted in a solution containing 10 mM Tris-HCl, pH 8.0, and 200 mM NaCl. Protein crystallization was performed using the batch method as previously described [19]. Equal volumes (20 μL each) of lysozyme (50 mg/mL) and crystallization solution containing 0.1 M Na-acetate (pH 4.5), 5% (w/v) PEG 6000, and 1 M NaCl were transferred into a 1.5-mL microtube and immediately vortexed for 30 s. The sample mixture was incubated at 4 °C. The protein crystals were grown under protein precipitation for two weeks. The crystal dimensions were 0.25 × 0.25 × 0.10 mm^3.

2.2. X-ray Diffraction Data Collection

The diffraction data were collected at Beamline 7A of the Pohang Accelerator Laboratory (Pohang, Korea). For native data collection, a single crystal of hLZM was soaked for 10 s in cryo-protectant solution (crystallization solution and 20% [v/v] ethylene glycol). The crystal was then mounted on the goniometer under a nitrogen gas stream at 100 K. For the hLZM-NAG complex, a single hLZM crystal was soaked for 30 min in a solution containing 0.1 M Na-acetate (pH 4.5), 5% (w/v) PEG 6000, 1 M NaCl, and 50 mM NAG. This crystal was then further soaked in a cryo-protectant solution and the diffraction data were collected. Indexing, integration, and processing of the diffraction data were performed using the HKL2000 program [20].

2.3. Structure Determination

The phases were obtained using the molecular replacement method with the MOLREP program [21]. The crystal structure of the chicken egg-white lysozyme (PDB ID: 7E02, sequence identity: 59%) [22] was used as the search model. The model was built using the program Coot [23]. Model refinement was performed using refine.phenix of the PHENIX program [24]. Data collection and refinement statistics are presented in Table 1. The geometry of the final models was validated using MolProbity [25]. The protein–ligand interaction was analyzed using PLIP [26]. Structure-based sequence alignments were generated using Clustal-Omega [27] and ESPript [28]. The structure was visualized using PyMOL (http://pymol.org).

Table 1. Data collection and refinement statistics.

Data Collection	**hLZM**	**hLZM-NAG1**	**hLZM-NAG2**
Wavelength (Å)	0.979	0.979	0.979
Space group	$P2_12_12_1$	$P2_12_12_1$	$P2_12_12_1$
Unit cell dimension (Å)			
a	33.035	32.796	32.980
b	56.062	55.758	56.049
c	61.063	60.859	60.939
Resolution range (Å) [a]	50.0–1.30(1.32–1.30)	50.0–1.55 (1.58–1.55)	50.0–1.60 (1.63–1.60)
Unique reflections	28,633 (1358)	16,579 (797)	15,319 (688)
Completeness (%)	99.5 (94.8)	99.6 (98.2)	99.3 (90.3)
Redundancy	6.4 (4.4)	9.5 (5.0)	6.0 (4.6)
Mean $I/\sigma(I)$	40.42 (4.53)	23.20 (3.37)	27.59 (3.23)
R_{merge}	0.095 (0.428)	0.293 (1.637)	0.108 (0.524)
CC1/2	0.989 (0.863)	0.981 (0.502)	0.993 (0.832)
CC*	0.997 (0.963)	0.995 (0.818)	0.998 (0.953)
Refinement			
Resolution range (Å)	50.0–1.30	50.0–1.55	50.0–1.60
	(1.33–1.30)	(1.60–1.55)	(1.65–1.60)
R_{work}	0.1513 (0.2055)	0.1686 (0.2236)	0.1683 (0.2034)
R_{free}	0.1710 (0.2056)	0.1910 (0.2664)	0.1844 (0.2572)
RMS deviations			
Bonds (Å)	0.022	0.009	0.006
Angles (degree)	1.639	1.146	0.902
Average B factor ($Å^2$)			
Protein	12.30	15.19	17.95
Ligand		26.48	32.57
Water	24.01	25.01	27.55
Ramachandran plot			
Most favored (%)	97.66	96.09	96.88
Allowed (%)	2.34	3.91	3.12

[a] Values in parentheses are for outer shells.

3. Results and Discussion

3.1. Overall Structure

All hLZM crystals belonged to the orthorhombic space group $P2_12_12_1$, occupying one molecule in an asymmetric unit. The crystal structures of a native and two NAG-bound hLZM crystals were determined at 1.30 and 1.55/1.60 Å resolution, respectively. The electron density was very clear for most of the residues between Lys19 and Val148 of all hLZM proteins, but not for Arg59–Ala60 of the native hLYZ. Validation of the geometry showed that >96% of the hLZM residues were within the most favored region of the Ramachandran plot and without outlier residues (Table 1). As hLZM contains eight cysteine residues that form disulfide bonds (Cys24–Cys146, Cys48–Cys134, Cys83–Cys99, and Cys95–Cys113), the lack of a significant negative Fo-Fc electron-density map indicated that all the hLZM structures determined in the current study were in an oxidized state. The hLZM was composed of seven α-helices and two β-strands, forming one α-domain and one β-domain (Figure 1a). The approximate dimensions of hLZM were 30 × 30 × 45 Å (Figure 1a). The active site of hLZM can accommodate six consecutive sugars at subsites A–F [2]. The catalytic residues Glu53 and Asp71 cleave the bond between subsites D and E (Figure 1b). The electrostatics of the active site of hLZM revealed a strong negatively charged surface (Figure 1b). The acetate ion of native hLZM was observed at subsite C and the α-D-NAG molecule of the hLZM-NAG complex was observed in the middle of subsites C and D (see below).

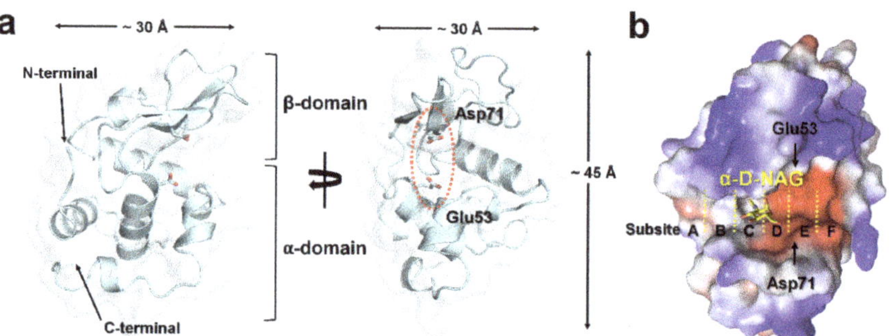

Figure 1. Crystal structure of hLZM. (a) Cartoon and surface representation of hLZM. **(b)** Electrostatics of the hLZM-NAG surface. Subsites in the active-site cleft of hLZM are indicated by the yellow dotted lines.

3.2. NAG Binding

To investigate the hLZM recognition of the peptidoglycan component, NAG ligand-soaking experiments were performed with hLZM and two crystal structures of NAG-bound hLZM were determined. In the previous peptidoglycan cleavage model, NAG units of peptidoglycan were located at subsites A, C, and E, whereas NAM units were positioned at subsites B, D, and F (Figure 2a). Crystallographic results of the NAG-soaking experiment revealed the electron-density map corresponding NAG was only observed at the middle of the subsites C and D of hLZM (Figure 2b), with the other subsites being occupied by water molecules. The NAG reagent used in this experiment was a mixture of *N*-acetyl-α-D-glucosamine (α-D-NAG) and the *N*-acetyl-β-D-glucosamine (β-D-NAG). Analysis of the refinement results and electron-density map showed that α-D-NAG was bound to hLZM (Supplementary Figure S1). This indicated that hLZM preferred the interaction with α-D-NAG compared with that of β-D-NAG when a single NAG molecule was interacting hLZM. To confirm that the NAG molecule reproducibly binds to hLZM, two diffraction data sets of the hLZM crystal soaked with NAG were collected. In the two crystal structures of α-D-NAG-bound hLZM, the quality of the electron density for α-D-NAG and the binding site was almost identical, while the quality of the electron-density map for α-D-NAG surrounding water molecules was slightly different (Supplementary Figure S2). The superimposition of the two α-D-NAG -bound structures showed that the position of α-D-NAG was almost identical with the root-mean-square deviation (RMSD) analysis of 0.141 Å (Supplementary Figure S3). This indicated that a single α-D-NAG molecule was specifically bound in the middle of subsites C and D in hLZM. The binding of α-D-NAG in the two α-D-NAG-bound structures was analyzed using the crystal structure of hLZM-NAG1, which was superior to that of hLZM-NAG2 in terms of data quality and electron density clarity.

The electron-density map of *N*-acetyl and the ring of glucosamine were clearly observed in hLZM-NAG1, whereas the electron-density map of the C6-O6 region of glucosamine was weaker. The α-D-NAG molecule was stabilized by hydrophobic interactions (Val117, Ala126, and Trp127), hydrogen bonds (Asn64, Asn78, Ala126, and Val128), and water bridges (Tyr81, Trp82, Ala126, and Ala129). Specifically, the *N*-acetyl group of α-D-

NAG was located at the main chains between residues Asn78 and Ala128 (Figure 2c). The C8 atom of the *N*-acetyl group of α-D-NAG interacted with the CG1 atom of Val117, CB atom of Ala126, and CD2 atom of Trp127 by hydrophobic interactions at distances of 3.60, 3.97, and 3.70 Å, respectively. The O7 and N2 atoms of the *N*-acetyl group of α-D-NAG interacted with the amine group of Asn78 and the carbonyl group of Ala126 at distances of 2.92 Å and 3.16 Å, respectively. The glucosamine group of α-D-NAG was located at the side chains between residues Asn78 and Val128 (Figure 2c). The O1 atom of the glucosamine of α-D-NAG interacted with amino group of Val128 at a distance of 3.27 Å. The distance between the side chain of Asn78 and ring center of glucosamine of α-D-NAG was 4.53 Å. In addition, five water bridges between α-D-NAG and hLZM were observed at distances of 2.79–4.02 Å. More detail information regarding the distances between hLZM and α-D-NAG are listed in Table 2.

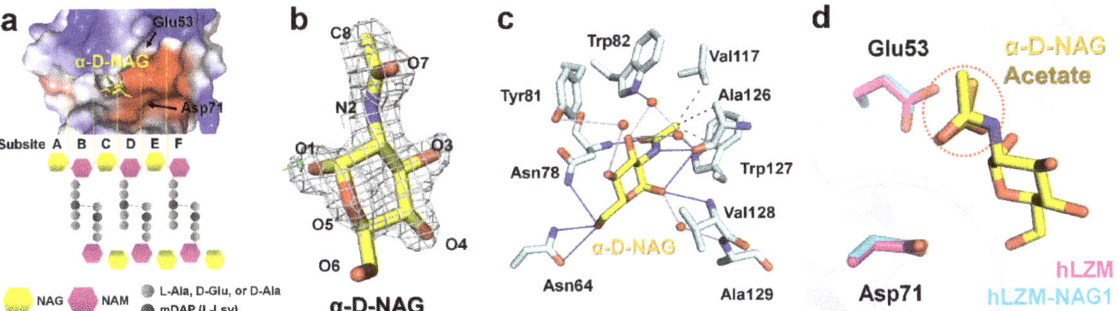

Figure 2. NAG binding to the active site of lysozyme. (**a**) Active-site cleft of hLZM and schematic representation of peptidoglycan. In peptidoglycan, *N*-acetylmuramic acid (NAM) is covalently linked to a peptide chain (2–5 amino acid), including a dibasic amino acid [typically meso-diaminopimelic acid (mDAP) or L-lysine]. (**b**) 2mFo-DFc (1.2 σ, grey mesh) and mFo-DFc (+3 σ: green mesh, −3 σ: red mesh) electron-density maps of α-D-NAG in hLZM-NAG. (**c**) Interaction between α-D-NAG and hLZM. Hydrophobic interactions, hydrogen bonds, and water bridges are indicated by grey dotted-lines, blue lines, and light blue lines, respectively. (**d**) Superimposition of native hLZM (magenta) and hLZM-NAG (cyan). The *N*-acetyl group of α-D-NAG (yellow) and acetate ion (orange) are indicated by the red dots.

Table 2. Interaction distances between hLZM and α-D-NAG.

Hydrophobic Interactions		
hLZM Residue [Atom]	**Distance (Å)**	**α-D-NAG Atom**
Val117 [CG1]	3.60	C8
Ala126 [CB]	3.97	C8
Trp127 [CD2]	3.70	C8
Hydrogen Bonds			
hLZM Residue [Atom]	**Distance (Å) H-A** [1]	**Distance (Å) D-A** [2]	**Donor Angle (°)**	**α-D-NAG Atom**
Asn64 [ND2]	2.06	2.92	145.86	O6
Asn64 [OD1]	2.17	3.16	177.97	O6
Asn78 [ND2]	2.31	3.27	164.54	O6
Asn78 [N]	2.90	3.57	127.25	O7

Table 2. *Cont.*

		Hydrogen Bonds		
hLZM Residue [Atom]	Distance (Å) H-A [1]	Distance (Å) D-A [2]	Donor Angle (°)	α-D-NAG Atom
Ala126 [O]	3.24	3.94	130.39	O1
Ala126 [O]	3.60	3.92	101.36	N2
Val128 [N]	3.79	4.08	100.14	O1

		Water Bridges			
hLZM Residue [Atom]	Distance (Å) A-W [3]	Distance (Å) D-W [4]	Donor Angle (°)	Water Angle (°)	α-D-NAG Atom
Tyr81 [OH]	4.00	3.48	127.96	112.25	O4
Trp82 [NE1]	3.02	2.72	144.92	87.42	O3
Ala126 [O]	2.79	3.71	102.23	107.74	O3
Ala126 [O]	4.02	2.72	144.92	114.21	O3
Ala129 [N]	3.01	3.23	158.53	77.42	O1

[1] Distance between hydrogen atoms and acceptor atoms. [2] Distance between donor and acceptor atoms. [3] Distance between acceptor and water atoms. [4] Distance between donor and water atoms.

The electron-density map corresponding to the acetate ion of native hLZM was observed at subsite C (Supplementary Figure S4). The oxygen atom interacted with the amino group of Asn78 at a distance of 2.90 Å (Supplementary Figure S5). The methyl group of acetate was stabilized by hydrophobic interactions with Val117 and Ala126 (Supplementary Figure S5), showing similarities with the binding mode of the *N*-acetyl group of α-D-NAG in hLZM-NAG. Superimposition of the native hLZM and hLZM-NAG structures showed that the position of the acetate ion in native hLZM was similar to that of the *N*-acetyl group of α-D-NAG in hLZM-NAG (Figure 2d).

3.3. Comparison of Native hLZM and NAG-Bound hLZM

To investigate how hLZM recognizes the α-D-NAG molecule, the structures of native hLZM and hLZM-NAG were compared. Superimposition of their crystal structures showed a high level of structural similarity with RMSD of 0.093–0.128 Å, indicating there was no large conformational changes upon α-D-NAG binding to hLZM; however, there was a small shift of the main-chain of the β1-strand and C-terminal helix (Figure 3a). The B-factor representation revealed that the β1-strand, α5-helix, and C-terminal α7-helix of α-D-NAG-bound hLZMs were more flexible compared with that of native hLZM (Figure 3b). Structural analysis of ligand binding revealed conformation changes of Asn62, Arg80, Tyr81, Val128, and Arg131 of hLZM upon α-D-NAG binding (Figure 3c). In particular, the two methyl groups of the side chain of Val128 of NAG-bound hLZMs were directed more toward the α-D-NAG molecule compared with that in native hLZM. The side chain of Val110 was separated from the ring center of glucosamine of α-D-NAG by 0.2 Å, indicating that α-D-NAG-binding induced conformational changes of the Val128 side chain. The side chain of Tyr81 in native hLZM was positioned toward subsite E, but the side chain of Tyr81 of α-D-NAG-bound hLZM was positioned toward the α-D-NAG molecule at subsite D, due to rotation of the side change by approximately 106° (Figure 3d). The Tyr81 residue of α-D-NAG-bound hLZM interacted with α-D-NAG via a water bridge (Figure 2c). The smallest distance between the glucosamine of α-D-NAG and phenol ring of Tyr81 was approximately 5 Å, whereas the distance between the ring center of glucosamine of α-D-NAG and the phenol ring of Tyr81 was approximately 7 Å in parallel dispatch. On the other hand, the CZ atoms of Arg80 and Arg131 were separated from α-D-NAG by 6.40 Å and 6.65 Å, respectively, resulting in no direct interaction between these atoms and α-D-NAG. It was considered that the sidechain shifts of Arg80 and Arg131 may have been induced by conformational changes of neighboring residues Tyr81 and Val128. Meanwhile, no significant conformational changes of active-site residues Glu53 or Asp71 were observed. Taken together, this suggests that amino acids present in the active-site cleft undergo conformational changes and stabilize α-D-NAG when α-D-NAG binds to hLZM.

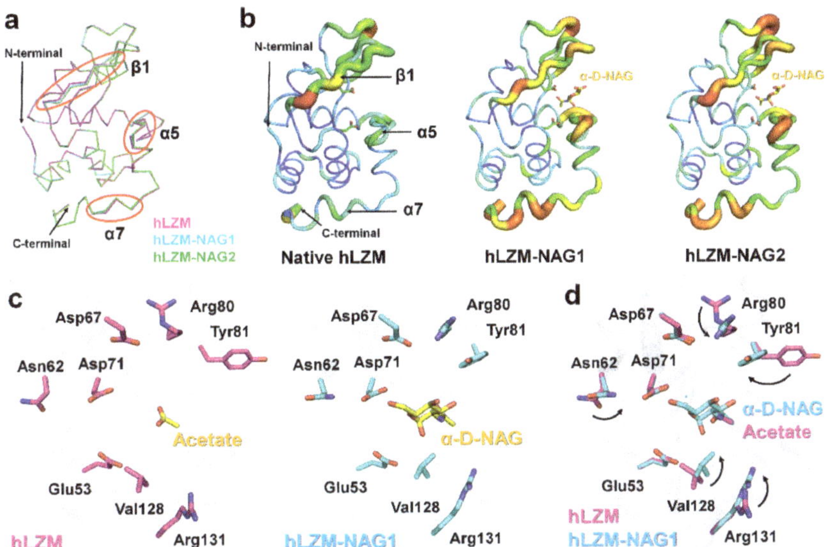

Figure 3. Comparison of native hLZM and α-D-NAG-bound hLZM. (**a**) Superimposition of the main chains of native hLZM (magenta), hLZM-NAG1 (cyan), and hLZM-NAG2 (green). (**b**) B-factor putty representation of native hLZM and α-D-NAG-bound hLZMs. (**c**) Ligand binding in native hLZM and hLZM-NAG1. (**d**) Superimposition of acetate and α-D-NAG binding from native hLZM and α-D-NAG-bound hLZM.

3.4. Structural Comparison of hLZM with Other Lysozyme Structures

To better understand the saccharide recognition of hLZM, the crystal structure of hLZM-NAG was compared with that of a previously reported crystal structure of hLZM complexed with a tetrasaccharide fragment of the O-chain of lipopolysaccharide from *Klebsiella pneumonia* (PDB ID: 5LSH). In this crystal structure, tetrasaccharide bound to the substrate-binding cleft of hLZM with two different conformations (Figure 4a), in which tetrasaccharide was located at the subsites A–D of hLZM. The superimposition of α-D-NAG-bound and tetrasaccharide-bound hLZM showed a high degree of structural similarity with RMSD of 0.216 Å, and two O6 atoms of tetrasaccharide demonstrated positional similarity with that of the *N*-acetyl group of α-D-NAG (Figure 4a). Meanwhile, the ring positions of α-D-NAG and tetrasaccharide were different due to differences in chemical structure (Figure 4a). Superimposition of these hLZM structures showed that most of the amino acids interacting with saccharide molecules underwent large conformational changes of their side chains on the active-site cleft of hLZM (Figure 4b). Residues Tyr81 and Val128, in particular, demonstrated large differences in side-chain conformation.

Subsequently, the crystal structure of hLZM-NAG was compared with those of hLZM complexed with 2′,3′-epoxypropyl α-glycoside derivatives of the disaccharides [NAG-NAG-EPO (PDB ID: 1REY) or GAL-NAG-EPO (PDB ID: 1REZ)]. In these complex structures, the NAG molecule is located on subsite C (Figure 4c), which coincides with the substrate binding site model. The second sugar of the two ligands (either NAG or GAL) is located on subsite B, but with a slightly different conformation (Figure 4c). This indicates that the sugar recognition mode of hLZM depends on the number and type of sugar. Superimposition of these structures showed that, in the active-site cleft of hLZM, most amino acids interacting with saccharide molecules underwent small conformational changes of their side chains (Figure 4d).

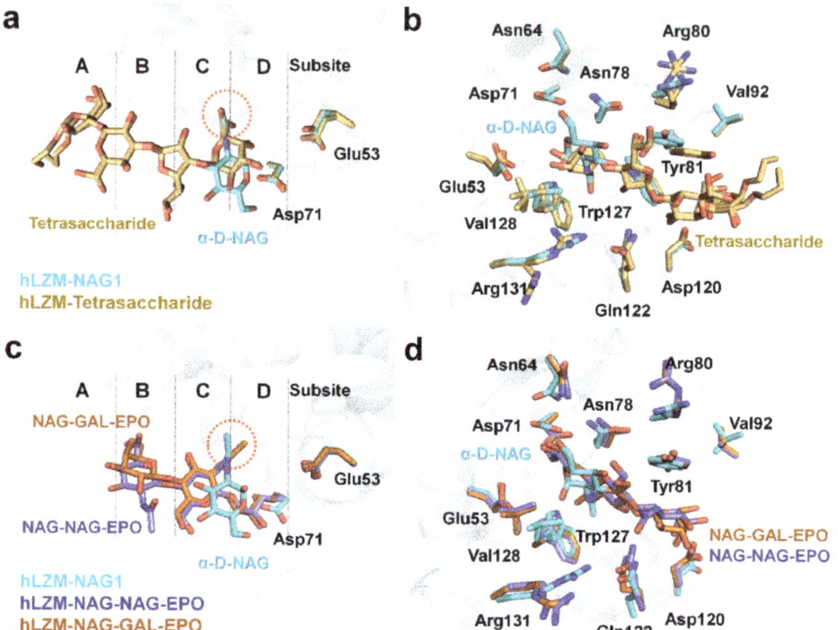

Figure 4. Comparison of α-D-NAG-bound and tetrasaccharide-bound hLZM. (**a**) Superimposition of hLZM-NAG (cyan) with a tetrasaccharide fragment from the O-chain of lipopolysaccharide from *Klebsiella pneumonia* (orange; PDB ID: 5LSH). The O6 atoms of tetrasaccharide and the *N*-acetyl group of α-D-NAG are indicated by red dots. (**b**) Superimposition of saccharide residues interacting with hLZM complexed with α-D-NAG (cyan) and the tetrasaccharide fragment (orange). (**c**) Superimposition of hLZM-NAG (cyan) with hLZM complexed with NAG-NAG-EPO (blue; PDB ID: 1REY) and NAG-GAL-EPO (brown; PDB ID: 1REZ). The *N*-acetyl groups of α-D-NAG, NAG-NAG-EPO, and NAG-GAL-EPO are indicated by red dots. (**d**) Superimposition of saccharide residues interacting with hLZM complexed with α-D-NAG (cyan), NAG-NAG-EPO (blue), and NAG-GAL-EPO (brown).

To evaluate the degree of conservation in the binding mode of α-D-NAG among the c-type lysozyme family, the crystal structure of hLZM-NAG was compared with that of the crystal structure of α-D-NAG-bound chicken egg-white lysozyme (cLZM), which is a well-studied model structure of c-type lysozyme. The antibacterial activity of hLZM is three-fold higher than that of the antibacterial activity of cLZM [1]. The sequence identity between hLZM and cLZM was 59% (Figure 5a). To date, three α-D-NAG-bound cLZM have been deposited in PDB. One of the crystal structures of cLZM (PDB ID: 1JA7) was determined in a powder diffraction study [29], while the other two crystal structure of cLZM (PDB ID: 7BHM and 7BHN) were determined by serial crystallography performed at room temperature [30]. In the case of the latter two crystal structures of α-D-NAG-bound cLZM, the results are highly similar due to only differences in the NAG mixing times being used. Accordingly, only one of these crystal structures of α-D-NAG-bound cLZM (7BHN) was compared in this study (the one that underwent the longer NAG-mixing time). Superimposition of the α-D-NAG-bound hLZM with that of the α-D-NAG-bound cLZM-1JA7 and cLZM-7BHN had an RMSD of 1.279 Å and 0.475 Å, respectively, for the main chain of hLZM. The glucosamine group of α-D-NAG was positioned toward to the active-site residues Asp35 and Glu71 in hLZM, whereas the *N*-acetyl group of α-D-NAG was positioned toward to active-site residues in cLZM-1JA7 (Figure 5b). As a result, a similarity in the α-D-NAG-binding modes for hLZM and cLZM-1JA7 was not observed. In contrast, in cLZM-7BHN, the glucosamine group of α-D-NAG was positioned toward the active site of cLZM, similar to that of α-D-NAG-bound hLZM (Figure 5c). Moreover, the binding mode of the *N*-acetyl group of α-D-NAG was similar between hLZM and cLZM (Figure 5c). Meanwhile, the position of the O6 atom in the glucosamine group of α-D-NAG from hLZM and cLZM differed by 3.95 Å.

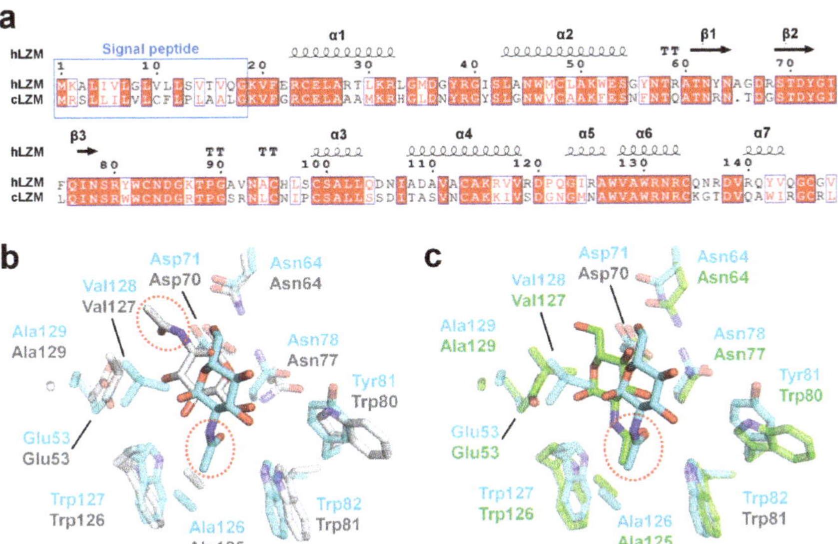

Figure 5. Comparison of hLZM and cLZM. (**a**) Sequence alignment of hLZM (Uniprot: P61626) and cLZM (P00698). (**b**) Superimposition of the crystal structures of α-D-NAG-bound hLZM with cLZM (PDB ID: 1JA7), which was determined through a powder diffraction study. (**c**) Superimposition of the α-D-NAG-binding sites of hLZM and cLZM (PDB ID: 7BHN), which was determined through serial femtosecond crystallography.

4. Conclusions

The hydrolases hLZM plays a significant role in the antibacterial and antiviral immune responses in human infants. In this study, the high-resolution crystal structure of hLZM complexed with *N*-acetyl-α-D-glucosamine, a component of peptidoglycan, was reported for the first time. Six subsites for sugar binding were identified in the cleft of the active site of hLZM, with the α-D-NAG molecule being only observed in the middle of subsites C and D. This result indicates that a single α-D-NAG molecule prefers to bind near the active site of hLZM. Although the binding mode of α-D-NAG to hLZM is not fully consistent with that of other hLZM-tetrasaccharide or cLZM-α-D-NAG binding modes, the position of saccharide binding was similar. This indicates that the specific NAG binding to hLZM and cLZM are distinct, but the subsites C and D are positionally important for the recognition of a single NAG molecule. Meanwhile, the previous peptidoglycan binding and cleavage model of hLZM suggested that the NAG molecule may prefers to interact with subsites A, C, or F; however, the current actual experimental results located α-D-NAG only in the middle of subsites C and D. This indicates that the modes of binding of a single α-D-NAG molecule compared with that of the actual substrate molecule consisting of NAG-NAM may differ. Accordingly, further structural and biochemical experiments will be required to identify the interactions between hLZM and their substrates. Nevertheless, this study provided a greater understanding of how hLZM recognizes single α-D-NAG molecules, which is a component of peptidoglycan, and how it differentiates it from other sugar-binding LZM structures. The results from this study will provide new insights into sugar recognition of hLZM and the role hLZM plays in the immune system.

Supplementary Materials: The following supporting information can be downloaded at: https://www.mdpi.com/article/10.3390/app12094363/s1, Figure S1: Electron-density map of hLZM after adding NAG and refinement, Figure S2: Electron-density map of hLZM-NAG1 and hLZM-NAG2, Figure S3: Superimposition of crystal structures of hLZM-NAG1 and hLZM-NAG2, Figure S4: Electron-density map of the acetate ion in native hLZM, Figure S5: Interaction between the acetate ion and hLZM.

Funding: This work was funded by the National Research Foundation of Korea (NRF) (NRF-2017M3A9F6029736 and NRF-2021R1I1A1A01050838) and the Korea Initiative for Fostering University of Research and Innovation (KIURI) Program of the NRF (NRF-2020M3H1A1075314).

Institutional Review Board Statement: Not applicable.

Informed Consent Statement: Not applicable.

Data Availability Statement: The coordinates and structure factors for the protein crystals of this study are available in the Protein Data Bank (https://www.rcsb.org/) under entries 7XF6 (hLZM), 7XF7 (hLZM-NAG1), and 7XF8 (hLZM-NAG2).

Acknowledgments: I thank the staff at the 7A beamline at Pohang Accelerator Laboratory for their assistance with data collection.

Conflicts of Interest: The author declares no conflict of interest.

References

1. Ferraboschi, P.; Ciceri, S.; Grisenti, P. Applications of Lysozyme, an Innate Immune Defense Factor, as an Alternative Antibiotic. *Antibiotics* **2021**, *10*, 1534. [CrossRef]
2. Callewaert, L.; Michiels, C.W. Lysozymes in the animal kingdom. *J. Biosci.* **2010**, *35*, 127–160. [CrossRef]
3. Bliska, J.B.; Ragland, S.A.; Criss, A.K. From bacterial killing to immune modulation: Recent insights into the functions of lysozyme. *PLoS Pathog.* **2017**, *13*, e1006512. [CrossRef]
4. Jollès, P.; Jollès, J. What's new in lysozyme research? Always a model system, today as yesterday. *Mol. Cell. Biochem.* **1984**, *63*, 165–189. [CrossRef]
5. Beintema, J.J.; van Scheltinga, A.C.T. Plant lysozymes. In *Lysozymes: Model Enzymes in Biochemistry and Biology*; Experientia Supplementum; Springer: Berlin/Heidelberg, Germany, 1996; pp. 75–86. [CrossRef]
6. Fastrez, J. Phage lysozymes. In *Lysozymes: Model Enzymes in Biochemistry and Biology*; Experientia Supplementum; Springer: Berlin/Heidelberg, Germany, 1996; pp. 35–64. [CrossRef]
7. Uversky, V.N.; Wohlkönig, A.; Huet, J.; Looze, Y.; Wintjens, R. Structural Relationships in the Lysozyme Superfamily: Significant Evidence for Glycoside Hydrolase Signature Motifs. *PLoS ONE* **2010**, *5*, e15388. [CrossRef]
8. Ellison, R.T.; Giehl, T.J. Killing of gram-negative bacteria by lactoferrin and lysozyme. *J. Clin. Investig.* **1991**, *88*, 1080–1091. [CrossRef]
9. Oliver, W.T.; Wells, J.E. Lysozyme as an alternative to growth promoting antibiotics in swine production. *J. Anim. Sci. Biotechnol.* **2015**, *6*, 35. [CrossRef]
10. Tenovuo, J. Clinical applications of antimicrobial host proteins lactoperoxidase, lysozyme and lactoferrin in xerostomia: Efficacy and safety. *Oral Dis.* **2002**, *8*, 23–29. [CrossRef]
11. Masschalck, B.; Michiels, C.W. Antimicrobial Properties of Lysozyme in Relation to Foodborne Vegetative Bacteria. *Crit. Rev. Microbiol.* **2008**, *29*, 191–214. [CrossRef]
12. Lu, D.; Liu, S.; Ding, F.; Wang, H.; Li, J.; Li, L.; Dai, Y.; Li, N. Large-scale production of functional human lysozyme from marker-free transgenic cloned cows. *Sci. Rep.* **2016**, *6*, 22947. [CrossRef]
13. Travis, S.M.; Conway, B.-A.D.; Zabner, J.; Smith, J.J.; Anderson, N.N.; Singh, P.K.; Peter Greenberg, E.; Welsh, M.J. Activity of Abundant Antimicrobials of the Human Airway. *Am. J. Respir. Cell Mol. Biol.* **1999**, *20*, 872–879. [CrossRef]
14. Siwicki, A.K.; Klein, P.; Morand, M.; Kiczka, W.; Studnicka, M. Immunostimulatory effects of dimerized lysozyme (KLP-602) on the nonspecific defense mechanisms and protection against furunculosis in salmonids. *Vet. Immunol. Immunopathol.* **1998**, *61*, 369–378. [CrossRef]
15. Zhang, R.; Wu, L.; Eckert, T.; Burg-Roderfeld, M.; Rojas-Macias, M.A.; Lütteke, T.; Krylov, V.B.; Argunov, D.A.; Datta, A.; Markart, P.; et al. Lysozyme's lectin-like characteristics facilitates its immune defense function. *Q. Rev. Biophys.* **2017**, *50*, e9. [CrossRef] [PubMed]

16. Osserman, E.F.; Klockars, M.; Halper, J.; Fischel, R.E. Effects of Lysozyme on Normal and Transformed Mammalian Cells. *Nature* **1973**, *243*, 331–335. [CrossRef]

17. Lee-Huang, S.; Maiorov, V.; Huang, P.L.; Ng, A.; Lee, H.C.; Chang, Y.-T.; Kallenbach, N.; Huang, P.L.; Chen, H.-C. Structural and Functional Modeling of Human Lysozyme Reveals a Unique Nonapeptide, HL9, with Anti-HIV Activity. *Biochemistry* **2005**, *44*, 4648–4655. [CrossRef]

18. Song, H.; Inaka, K.; Maenaka, K.; Matsushima, M. Structural Changes of Active Site Cleft and Different Saccharide Binding Modes in Human Lysozyme Co-crystallized with Hexa-N-acetyl-chitohexaose at pH 4.0. *J. Mol. Biol.* **1994**, *244*, 522–540. [CrossRef]

19. Nam, K.H.; Cho, Y. Stable sample delivery in a viscous medium via a polyimide-based single-channel microfluidic chip for serial crystallography. *J. Appl. Crystallogr.* **2021**, *54*, 1081–1087. [CrossRef]

20. Otwinowski, Z.; Minor, W. Processing of X-ray diffraction data collected in oscillation mode. *Methods Enzymol.* **1997**, *276*, 307–326. [CrossRef]

21. Vagin, A.; Teplyakov, A. Molecular replacement with MOLREP. *Acta Crystallogr. D Biol. Crystallogr.* **2010**, *66*, 22–25. [CrossRef]

22. Nam, K.H. Beef tallow injection matrix for serial crystallography. *Sci. Rep.* **2022**, *12*, 694. [CrossRef]

23. Emsley, P.; Cowtan, K. Coot: Model-building tools for molecular graphics. *Acta Crystallogr. D Biol. Crystallogr.* **2004**, *60*, 2126–2132. [CrossRef]

24. Liebschner, D.; Afonine, P.V.; Baker, M.L.; Bunkoczi, G.; Chen, V.B.; Croll, T.I.; Hintze, B.; Hung, L.W.; Jain, S.; McCoy, A.J.; et al. Macromolecular structure determination using X-rays, neutrons and electrons: Recent developments in Phenix. *Acta Crystallogr. D Struct. Biol.* **2019**, *75*, 861–877. [CrossRef]

25. Williams, C.J.; Headd, J.J.; Moriarty, N.W.; Prisant, M.G.; Videau, L.L.; Deis, L.N.; Verma, V.; Keedy, D.A.; Hintze, B.J.; Chen, V.B.; et al. MolProbity: More and better reference data for improved all-atom structure validation. *Protein Sci.* **2018**, *27*, 293–315. [CrossRef]

26. Salentin, S.; Schreiber, S.; Haupt, V.J.; Adasme, M.F.; Schroeder, M. PLIP: Fully automated protein–ligand interaction profiler. *Nucleic Acids Res.* **2015**, *43*, W443–W447. [CrossRef]

27. Sievers, F.; Wilm, A.; Dineen, D.; Gibson, T.J.; Karplus, K.; Li, W.; Lopez, R.; McWilliam, H.; Remmert, M.; Soding, J.; et al. Fast, scalable generation of high-quality protein multiple sequence alignments using Clustal Omega. *Mol. Syst. Biol.* **2011**, *7*, 539. [CrossRef]

28. Gouet, P.; Courcelle, E.; Stuart, D.I.; Metoz, F. ESPript: Analysis of multiple sequence alignments in PostScript. *Bioinformatics* **1999**, *15*, 305–308. [CrossRef]

29. Von Dreele, R.B. Binding of *N*-acetylglucosamine to chicken egg lysozyme: A powder diffraction study. *Acta Crystallogr. D Biol.* **2001**, *57*, 1836–1842. [CrossRef] [PubMed]

30. Butryn, A.; Simon, P.S.; Aller, P.; Hinchliffe, P.; Massad, R.N.; Leen, G.; Tooke, C.L.; Bogacz, I.; Kim, I.-S.; Bhowmick, A.; et al. An on-demand, drop-on-drop method for studying enzyme catalysis by serial crystallography. *Nat. Commun.* **2021**, *12*, 4461. [CrossRef]

MDPI
St. Alban-Anlage 66
4052 Basel
Switzerland
Tel. +41 61 683 77 34
Fax +41 61 302 89 18
www.mdpi.com

Applied Sciences Editorial Office
E-mail: applsci@mdpi.com
www.mdpi.com/journal/applsci